COMPUTER CONTROL
IN THE PROCESS
INDUSTRIES

by Brian Roffel
Patrick Chin

CRC Press
Taylor & Francis Group
Boca Raton London New York

CRC Press is an imprint of the
Taylor & Francis Group, an **informa** business

First published 1987 by CRC Press
Taylor & Francis Group
6000 Broken Sound Parkway NW, Suite
300 Boca Raton, FL 33487-2742

Reissued 2018 by CRC Press

© 1987 by Taylor & Francis
CRC Press is an imprint of Taylor & Francis Group, an Informa business

No claim to original U.S. Government works

A Library of Congress record exists under LC control number: 87016913

Publisher's Note
The publisher has gone to great lengths to ensure the quality of this reprint but points out that some imperfections in the original copies may be apparent.

Disclaimer
The publisher has made every effort to trace copyright holders and welcomes correspondence from those they have been unable to contact.

ISBN 13: 978-1-138-50526-1 (hbk)
ISBN 13: 978-1-138-55790-1 (pbk)
ISBN 13: 978-1-315-15048-2 (ebk)

Visit the Taylor & Francis Web site at http://www.taylorandfrancis.com and the CRC Press Web site at http://www.crcpress.com

PREFACE

Techniques such as dead time compensation, adaptive control, and Kalman filtering have been around for some time, but as yet find little application in industry. This is due to several reasons, including:

- Articles in the literature usually assume that the reader is familiar with a specific topic and are therefore often difficult for the practicing control engineer to comprehend.
- Many practicing control engineers in the process industry have a chemical engineering background and did not receive a control engineering education.
- There is a wide gap between theory and practical implementation, since implementation is primarily concerned with robustness, and theory is not. The user therefore has to build an "expert shell" in order to achieve the desired robustness. Little is published on this issue, however.

This book tries to promote the use of advanced control techniques by taking the reader from basic theory to practical implementation. It is therefore of interest to practicing control engineers in various types of industries, especially the process industry. Graduate and undergraduate students in control engineering will also find the book extremely useful, since many practical details are given which are usually omitted in books on control engineering.

Of special interest are the simulation examples, illustrating the application of various control techniques. The examples are available on a 5-1/4″ floppy disk and can be used by anyone who has access to LOTUS 1-2-3.

Chapter 1 is the introduction; Chapters 2 through 6 deal with distributed control system networks, computer system software, computer system selection, reliability and security, and batch and continuous control. Chapter 7 gives an introduction to advanced control. Chapters 8 through 11 deal with dead time compensation techniques and model identification. Chapters 12 through 14 discuss constraint control and design, and the adjustment and application of simple process models and optimization. Chapter 15 gives a thorough introduction to adaptive control, and the last two chapters deal with state and parameter estimation.

This book is a valuable tool for everyone who realizes the importance of advanced control in achieving improved plant performance. It will take the reader from theory to practical implementation.

ABOUT THE AUTHORS

Brian Roffel is Global Advisor, Advanced Computer Control at Polysar Limited. He holds a M.Eng. in chemical engineering and a Ph.D. in process control. He worked as chief senior research officer at the Twente University of Technology, and as a computer applications engineer at Esso Chemical Canada. During 1986–87 he was adjunct professor at the Universities of Western Ontario and Waterloo. He has written many publications on process dynamics and control.

Patrick A. Chin is a Computer Control Specialist at Polysar Limited. He holds a M.Eng. in chemical engineering from McMaster University with a specialization in process control. Previously he worked as computer applications engineer at Esso Chemical Canada.

Contents

ix

List of Figures

List of Tables

CHAPTER 1

Introduction

Computers play an integral part in daily life. With the introduction of the chip they now are within the reach of almost everyone. Computers are being increasingly used in education, which has resulted in much greater familiarity with concepts that were known only to computer specialists 15 years ago. Nevertheless, much misconception exists regarding what a computer can do. Many people think that computers can perform miracles, but efficient software is needed for each specific task that has to be performed. There is also the tendency to use a computer in the manner that a pocket calculator has been used, that is: without knowing how it works you can still use it for many purposes.

One of the problems in many companies today is that managers are not always aware that computers for plant monitoring and control cannot be compared to the IBM-PCs on their desks. The purposes of this book are to illustrate that the introduction of computers in a plant requires much more than the purchase of hardware and software and to introduce some basic concepts that everyone who is involved in plant automation should know.

PRODUCTIVITY VS FLEXIBILITY

Computers are essential if one wants to achieve increased product quality and production flexibility in order to remain competitive. The two key words in today's competitive environment are quality and flexibility. This is also true for the chemical and petrochemical industry. Ten years ago operations were oriented toward production. High volumes were processed at relatively low cost. Even though product quality was not always good, it was less important than volume. Besides, customers did not have the equipment to analyze the product they bought and depended entirely on data provided by the product manufacturer for quality analysis.

However, more sensitive analytical equipment became available at lower prices, enabling many customers to test the materials they purchase. Sometimes, by using statistical quality control techniques, they can even determine if they buy material which has been blended with other material to meet quality requirements.

The next decade will be characterized by decreased demand for products but higher specifications for product quality. The chemical and petrochemical process industry in particular will have to adjust to this situation by producing a larger variety of products at lower volumes with increased quality. To achieve these results effectively and economically, in-plant computers will be necessary for monitoring and improved control.

PROBLEMS IN PLANT AUTOMATION

The introduction of a process computer into a plant for monitoring and control can produce a number of problems. First, financial limitations have to be overcome. Although personal computers are continually decreasing in price, process computers are not. This is probably because they are different from any other type of computer: they are tied to a real process and therefore require extensive checking of hardware and software to make sure that the right outputs are transmitted to the field. Although the hardware may become less expensive, system software becomes more complicated in order to give increased system functionality, and this largely offsets the reduced hardware cost. For example, ten years ago process computers usually communicated with a limited number of black and white consoles. Today's systems support multiple-color operator stations.

The second problem is lack of knowledge. To control a process effectively an in-depth understanding of the process is required. Although some technology is reasonably well developed, e.g., olefins technology, there are many chemical processes which are not well understood. Operators run the plant by experience, but the computer does not have this prior knowledge base. It is important that what has to be controlled is well defined.

In order to develop a system that controls and monitors the plant, experienced personnel who are capable of translating functional specifications for process operation into computer software are required.

The computer system is usually so complicated that personnel are required to maintain and improve the computer software, whereas other specialists develop and maintain the software that runs the plant (application software). Although system software comes with the computer, application software needs to be developed because it is specific for each plant. This often requires a large number of person-years, which is often underestimated.

Another problem is that processes can change. Therefore, the resources have to be available to modify the software in order to deal with the new situation. In these cases, especially, one often finds that software that was developed some time ago is poorly documented or not documented at all. This can lead to a waste of many person-years, and a small change may require as much time as the original design.

ORGANIZATIONAL CHANGE

The introduction of a computer system into a plant must be accompanied by a change in organization. System and application engineers will become a resident part of the plant operating personnel. The system and application engineer supervisor will have a responsible task with a wide range of activities. He must be a technical manager with enough knowledge to stimulate and support the engineers. He will have to keep abreast of new technologies in order to make sure that they can be used whenever the need arises.

On the other hand, he has to relate to process engineers, supervisors, and managers in order to create a team environment in which his engineers can best function. This is often not well understood. In many companies process engineers list their desires for plant automation and expect the application engineers to do the job. Many process supervisors and managers do not understand what is involved in using process computers and underestimate the effort required to design and implement a control, monitoring, and optimization package. Proper decisions can be made only in a team environment.

COST ESTIMATE

Recognizing that plant automation may produce some problems, it is logical to list them all and attack them one by one. After the decisions have been made on organizational change, technical support, and operations support, the computer system requirements for control, monitoring, and information processing should be defined. Once the needs are defined, one can ask for bids from computer vendors. Chapter 6 provides assistance in computer selection.

It is important to realize that the purchase of a computer will generate a number of other costs. An example of the distribution of costs for automation of a chemical plant is given in Giles and Bulloch [1] and shown in Table 1.1. Although these numbers will vary from plant to plant, they clearly indicate that computer hardware and software are only a small portion of the total cost of plant automation. A process computer makes plant automation more feasible than before, but often additional instruments, especially analytical instruments, must be purchased and installed to fully utilize the increased capability.

Table 1.1.   Distribution of Costs for a Process Computer System

	Cost (%)
Computer hardware	13%
Process interface	24%
Analytical instruments	11%
Software	18%
Instrumentation engineering	8%
Application engineering	20%

MODERN COMPUTER SYSTEMS

Modern computer systems are characterized by distributed data processing. Figure 1.1 shows the structure of a local plant automation system. It may be connected to a plantwide information and management system that collects data from the various plants and stores it for further processing. The local system can also be connected to local analytical instruments, which usually have their own microcomputer.

The computer system collects data from the plant and processes it for monitoring and control purposes. Signals are converted via D/A and A/D converters, digital to analog, and analog to digital converters.

Usually two data channels are used for communication in order to increase the total system reliability. As will be discussed in a later chapter, data channels are the most critical part of modern computer systems.

There are a number of tasks for the computer system, as shown in Figure 1.2 [2]. Although these tasks will be dealt with in detail in a later chapter, they will be briefly reviewed here.

Planning is the highest level of activity. At this level the distribution of raw materials and products is controlled. The process plants have to be operated in such a way that inventories are maintained. When there are many plants involved, a complex problem is created, which is often solved off-line on a mainframe computer. There is an increasing tendency, however, to perform these activities on an on-line plant management system, which is connected to all process computers.

The next level of activity is control of the mode of operation. This indicates that during a certain period, certain products must be made out of certain raw materials. Each mode of operation can be improved by applying optimization of throughput, and so on. This activity can be done in the process control computer; the optimization of large and complex plants is often performed by a separate computer. The result of the optimization is the calculation of setpoints (desired values) for the lower level of control, e.g., impurities in distillation tower overhead and bottom compositions.

Next is the level where quality control takes place. A large part of this book

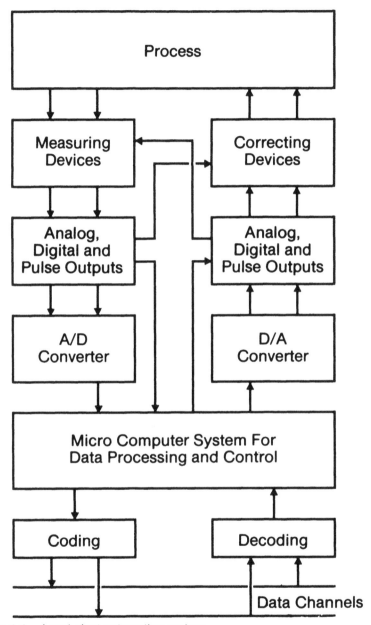

Figure 1.1. Local plant automation system.

will be dedicated to it. Quality control in the chemical and petrochemical industry creates some special control problems which have never received sufficient attention in the literature. Although process analyzers for quality

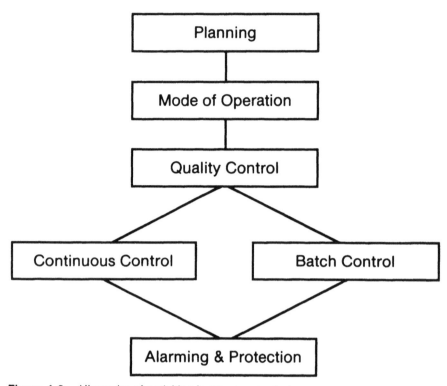

Figure 1.2.   Hierarchy of activities in process control.

control are still fairly expensive, they have to be used in order to produce high quality products. In this book some techniques will be discussed to accomplish quality control in a different way, although an analyzer can never be replaced by any of these techniques.

The next level of activity is continuous and batch control. In continuous control, process values are maintained close to their targets despite process upsets, equipment fouling, cleaning, and so forth. One should never try to use the computer as a replacement for conventional controllers. Conventional controllers today are modern microcomputers, which can do an excellent job. The computer should be used to do more advanced control (e.g., adaptive control or control using Kalman filters).

In batch process control, pumps have to be started and stopped, and valves have to be opened and closed in a certain sequence. Computers interfacing to programmable controllers or batch controllers can provide tremendous savings in batch process operation. Unfortunately batch control design is still much of an art: no methodology or optimal design procedure yet exists.

The lowest level of activity involves alarming and protection, although this is often done in multiple microprocessors. The process computer can play a significant role, however, by informing the operator of possible causes of problems and by presenting selective information.

COMPUTER HARDWARE AND SOFTWARE

The heart of a computer is a central processing unit with main memory (Figure 1.3). The main memory is usually a combination of random access and read-only memory. The central processing unit can be considered as an instruction set, ready to manipulate data. Data is stored in main memory together with the program for use of the instruction set. This general set-up of the computer is one of the reasons for the low price of the computer hardware. Any specialist implementing computers in industry should have a reasonable understanding of computer hardware and software layout. Chapters 2 and 3 are dedicated to this topic.

Another important part of the computer system is the peripherals through which the user can communicate with the system. Color CRTs are used for presentation of data, graphic information, and trends. For system software development, black and white consoles are preferable, however.

Although alarm printers were popular many years ago, most process alarming today is done through the color CRTs in order to avoid the noise of the alarm printers. Alarms are stored in an alarm file of which specific parts can be printed on request.

For software development, a high-speed printer is essential.

Disks are increasingly being replaced by solid state electronic memory, mainly because of the higher reliability, although disks remain important for massive data storage.

For communication, the computer uses a data channel or highway. For short distances, one can use a parallel data channel. In this case, data is transported along many parallel channels. Usually, however, when distances are long, the serial data channel is used. This is a coaxial or glass fiber cable along which

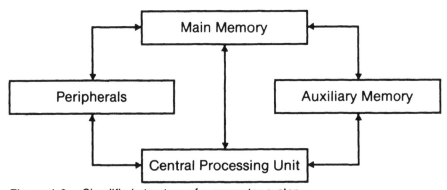

Figure 1.3. Simplified structure of a computer system.

information is passed in sequence. The rate of information processing is obviously lower than with a parallel data channel.

Computer system software is often classified into two categories: system software and application software. The system software provides general functions, such as compilation of programs, linking of programs, communication with peripherals, mathematical manipulations, data collection and storage, text editing, and servicing of console functions. A detailed description is given in Chapter 3.

Application software is the software that is used for monitoring and control. To use this software, the user often has to learn a language specific to the computer system. With this software complicated control strategies can be designed. Often the computer system also offers a higher level language, e.g., Pascal or Fortran. This is especially important when one wants to write one's own software, e.g., to optimize a section of the plant, to produce custom-designed reports or perform functions which are not standard on the system. The ability to use a higher level language may play an important part in system selection.

APPLICATION SOFTWARE DESIGN

As mentioned before, application software is the software that is used for monitoring and control. The design of this software is not an activity that should be done by a computer application engineer alone. Careful planning and teamwork are required.

First, standards for display building, program design, and documentation should be written and adhered to. In order to build process displays and to design control, monitoring, or optimization programs, the input from plant operators and plant engineers is desired. It is bad practice to present a complete system to the operators; if they input into various designs or even do part of certain designs (e.g., build displays), they will take ownership of the new computer system rather than having it forced upon them.

Application design should involve a continuous dialogue between the application engineer(s) and the process engineer(s). Only a team effort can make an automation project a success.

SUMMARY OF FOLLOWING CHAPTERS

The contents of this book are strongly oriented toward the engineer in the field who is using or is going to use and apply process computers. Chapters 2 through 5 are basic — things that every chemical and control engineer should know about process computers before using them. Chapter 2 deals with computer hardware, Chapter 3 with computer software, and Chapter 4 with computer selection. This chapter is particularly important for first-time users. Chapter 5 discusses reliability and security. Chapter 6 deals with batch and continuous control.

The following chapters focus more on the practical use of computers in implementing control strategies. Chapter 7 gives an introduction to advanced control, describing some simple process models and control elements. Difference equations and z-transform are introduced as a requirement for the next chapters.

The following chapters deal with techniques applied in quality control. Many chemical and petrochemical processes involve large dead times and time lags. Techniques are presented to deal with these situations. Many practical guidelines are given. Chapter 12 deals with constraint control as a means of plant optimization. Chapter 13 discusses the design, adjustment, and application of simple process models, whereas Chapter 14 focuses on optimization. The following three chapters discuss more advanced control techniques, such as adaptive control and Kalman filters.

CHAPTER 2

Distributed Computer Control System Networks

COMPUTER HARDWARE

The purpose of this chapter is to make the reader familiar with the concepts of modern computer networks. It is possible to discuss the advantages and disadvantages of different network structures without having a knowledge of computer hardware; however, it will be to the reader's benefit to have a basic understanding of the functional operation of a computer.

The four basic elements of a computer are the input/output circuit, the central processing unit, the (fast) main memory, and the (slower) auxiliary memory. This structure, called a Von Neumann structure, is usually applied in general purpose computers (Figure 2.1).

Process computers sometimes employ the Harvard structure (Figure 2.2). In this set-up there is another read-only memory (ROM), in which standard programs are stored, which can perform dedicated functions.

The central processing unit can be divided into an arithmetic and logic unit (ALU) and a control unit. The ALU performs the actual data manipulations that are required and will use temporary storage locations (registers) to store

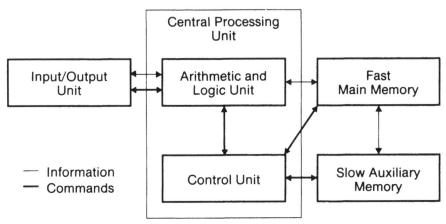

Figure 2.1. Von Neumann computer structure.

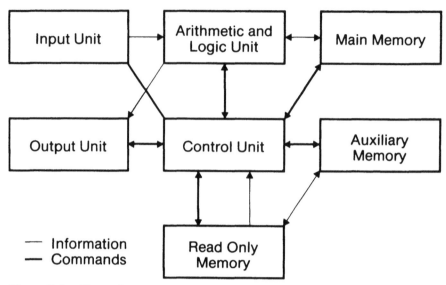

Figure 2.2. Harvard computer structure.

its data. Each register can store a binary word (which is usually equal to 8, 16, or 32 bits). The registers which determine which parts of the ALU will be used and how it will be used are called the control unit. Some of the operations that can be performed by the ALU are:[3]

- transfer between registers (move)

- transfer between registers and main memory (load and store)

- general register operations, such as clear, increment, or decrement

- elementary logical operations, e.g., "and" or "or"

- rotation and shift operations

- elementary manipulation with integer and floating point numbers

- stack operations

- the testing of the status of peripherals

Another important basic element of a computer is its memory. A distinction is made here between main memory and auxiliary memory. Main memory is used to store data and instructions that are used in the execution of a program. The computer main memory is random access (RAM), and the access time is short.

Final and longer term storage of data occurs in the auxiliary memory. The auxiliary memory is characterized by large storage capacity and relatively long access times. This memory is of the nonvolatile type, which means that data is not lost during a power outage. Some examples of this type of memory are: magnetic tapes, magnetic disks, magnetic bubble memories, and special solid state electronic memories. Boullart et al.[3] give a summary with access times and storage capacities; the results are summarized in Figure 2.3. In some computer systems a cache memory, a small buffer memory between the auxiliary memory and the central processing unit, is used. By filling this cache memory with information from the auxiliary memory, more rapid retrieval is possible.

The computer control unit executes instructions in the right sequence, interprets each instruction, and applies the right commands in the arithmetic unit and other logical units in accordance with this interpretation. The control unit determines the operation that has to be performed for addressing the main computer memory. The basic elements in the computer must communicate via a so-called data highway. Also, the user must communicate through the appropriate peripherals with the computer system. In early computer systems the central processing unit was considered to be the heart of the computer, and systems had the typical configuration shown in Figure 2.4. Recently, however, the trend has been toward distributed computer systems with the data highway

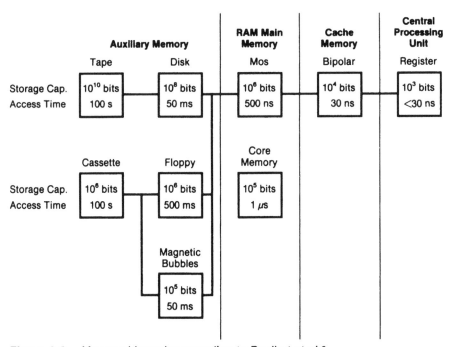

Figure 2.3.   Memory hierarchy according to Boullart et al.[3]

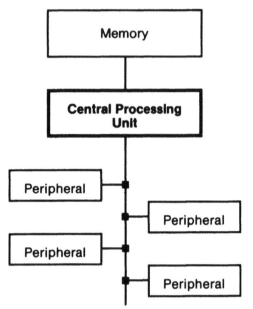

Figure 2.4.   Central processing–oriented communications structure.

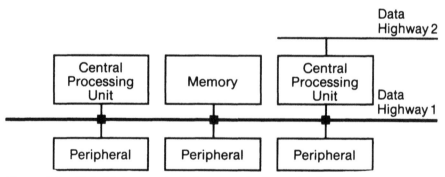

Figure 2.5.   Data highway–oriented communications structure.

as heart of the system. The central processing unit, peripherals, and memory are connected to the highway as shown in Figure 2.5.

PERIPHERALS

So far only the computer central processing unit, the main memory, and the auxiliary memory have been discussed. The fourth basic element, as previously mentioned, was the input/output unit. This unit is responsible for establishing communication between the computer and its environment, i.e., human being or process-oriented environment.

Keyboards, printers, CRTs, and so forth are used in communicating with process operators/engineers. In communicating with the process various indicator (measurements, switches) signals have to be converted to a form that can be processed by the computer. For example, analog temperature readings have to be converted into a binary number (A/D converter). The output signal to a control valve, represented by a binary number in the computer, has to be converted to an analog field signal (D/A converter). Very often field instrument systems have their own built-in microprocessor for signal conditioning.

The physical connection between a computer and a peripheral is established by an interface. This interface contains all hardware to communicate with the central processing unit and the specific peripheral. The interface will be dependent on the internal organization and construction of the computer, hence an interface will be computer/peripheral dependent. The physical connection between computer and peripheral can be realized in many different ways. The realization has an important impact on the programming techniques.[2] For example, it is possible to configure the computer system in such a way that data between computer memory and the peripheral can be exchanged only via the central processor, or in such a way that certain peripherals have direct memory access (DMA).

With the increase in complexity, speed, and capacity, the interface will become increasingly specific. There are two general systems: the parallel and serial interface. In a parallel interface, signal transfer is accomplished via a number of parallel lines, while other lines have a control function. Data transfer is synchronized by means of "handshaking" between the parties involved. The rate of transfer is high but becomes expensive at large distances, in which case a serial interface is preferred. In a serial interface, information bits are sent consecutively over one line, so the sender and receiver also have to be synchronized. Synchronization can be done by asynchronous transmission in which control bits are added to the transferred information; in synchronous transmission, transmitter and receiver have a common clock. Asynchronous transmission is easy to implement; synchronous transmission, however, is much faster.

A wide variety of mechanisms is used at all levels to provide the functions of an input/output system. There are two ways the central processor can become aware of an event: by interrupts and by polling. For interrupts, an interrupt handler has to be written and placed in the primary memory location to which the interrupt would cause a transfer. In polling, the central processing unit regularly checks a hardware register to determine whether or not a flag has been set. In slow peripherals much time will be wasted with polling; a more efficient way of I/O transfer is via interrupts. Special hardware and software are required to generate and handle the interrupts correctly. The most efficient way is the so-called vectored interrupt, where the peripheral regenerates an interrupt and then informs the central processor of its identity.[3-5] Some peripherals that service in communication with the computer are given in Table 2.1. In the process industry, high-speed printers, CRTs, and floppy disks have become familiar input/output devices.

DISTRIBUTED INPUT/OUTPUT HANDLING

In recent years a trend has developed toward the use of distributed data processing. Functions handled by the computer processing unit are now taken over by a subsystem with its own central processing unit that handles all the activities of the subsystem. The advantage is clearly that the main central processing unit is unloaded and can perform the dedicated data processing function for which it was designed. The main central processing unit is assisted by a processor that handles all the terminals, another processor handles the data on bulk storage devices, a third processor handles CRTs, and so forth. The structure can be organized in such a way that certain processors have direct memory access. A typical example of the described approach is shown in Figure 2.6.

MAJOR CONTROL SYSTEM FUNCTIONS

In the previous sections the basic structure of a computer system was discussed. The four main components of a computer, namely, the input/output system, the central processing unit, the main memory, and the auxiliary memory were reviewed. It was shown that there is a tendency toward distributed systems based on the data highway. Before discussing the structure of a computer control system, it is worthwhile to summarize the functions the computer has to perform. This will lead directly to the distributed control system configuration discussed in the next section.

Table 2.1.   Some Input/Output Devices

DEVICE	Typical Speed (Characters/sec)
Teletype	10
Character printer	60
Optical character reader	400
Video terminal	1,200
Line printer	1,300
Graphics terminal	19,200
Magnetic tape	100,000
Magnetic disk	300,000

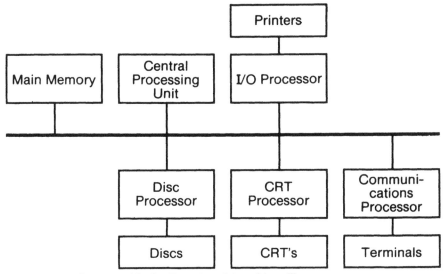

Figure 2.6.   System with distributed peripheral handling.

Early computer systems were used mainly for plant monitoring. Analog process signals were sampled, digitized, and conditioned. The input signals were converted into engineering units and alarm-checked, and then processing was done as defined in the database.

Control algorithms

When computers started to be used for control purposes, control algorithms were added to the database. General system software was usually available to perform the particular function, and the user could define his control structure and algorithm via a "fill in the blanks" method. Control algorithms vary considerably in complexity from simple feedback control to advanced control (e.g., adaptive control, multivariable control, Kalman filter state and parameter estimation). Most modern systems provide standard control functions such as feedback, feed-forward, and ratio control. However, implementation of advanced control is often left to the user's imagination and expertise.

Other control functions

Computer control systems can also provide safety interlocks to prevent equipment damage and protect the environment against human error. More recent systems often have software packages designed for plant optimization. Setpoints are not held at fixed values but are moved with changing operating conditions such that a predefined objective function is optimized. In batch control, the computer is used for sequential operation of valves, pumps, and so forth. It is obvious that there should be an excellent interface between the continuous and batch control packages in order to broaden the scope of possible applications. Another important task of the control system is to make all process information available to the operator. Other tasks will be described in a later chapter.

All previously mentioned tasks are susceptible to error, hence the impact of errors should be minimized. This can be accomplished by organizing the computer control system in a hierarchical modular fashion, so that when one module fails, other modules continue to operate as desired. The communications network plays a central role in these distributed systems.

DISTRIBUTED CONTROL MODULES AND NODES

The control functions discussed in the previous section are now organized in separate physical modules. Modules with similar functions are grouped together in physical structures called nodes[6,7] (see Figure 2.7). The nodes are linked to the central communications network by coaxial cable. Within a control node there could be an input module, an output module, and a control

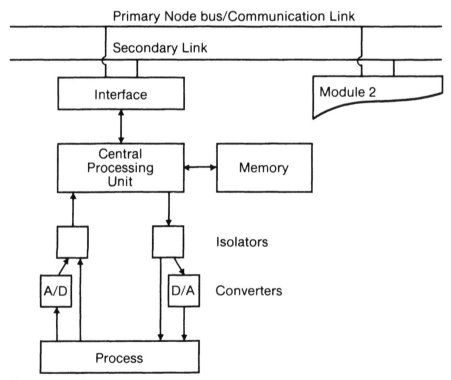

Figure 2.7.   General purpose control module (GPCM).

module, each module having its own microprocessor with error detection and recovery. A node bus interface links the node to the data highway. The distributed control node approach provides us with high application flexibility, easy handling of backup systems, and error recovery.

Even a data highway failure will not have a direct impact on the operation of a local node. The network must support a variable configuration and often operate under heavy loading since large amounts of information have to be transferred. This leads to specialized data highway communication networks.[7,8] The major types are the ring-based, broadcast-based, and cluster-based systems.

In ring-based systems, a node will transmit a message that enters the communications ring. Upon reaching its destination, the receiving node will act upon the message and pass the message on with a positive or negative acknowledgment. The message enters the ring again, and, when it is returned to the originating node, it is taken out of the ring. Each node maintains a buffer where incoming and outgoing messages are queued. When a message with negative acknowledgement is received, it is retransmitted. Experience with this type of system indicates that delays do not increase very much as a result of increased network loading.

In broadcast-based communication systems, each node contains a functional processor, a data highway controller to interface the node to the rest of the system, and a shared random access memory. The functional processor is customized through ROM software and special hardware. Each critical node has a redundant combination of the data highway controller, shared memory, and functional processor. The data highway consists of a multidrop coaxial cable. A token passing protocol is used in passing control from one module to another. Communication is organized around a 0.1-second cycle. During the first part of the cycle, the node broadcasts a 16-byte message containing database information and then passes the token on. Broadcast messages have no specific destination nor are they acknowledged. A typical system can broadcast around 10,000 points per second. The second half of the 0.1-second cycle can be used for other communications, e.g., changing controller parameters. The data highway controller formulates the messages, is responsible for the token passing protocol, and monitors the data highway.

A third data highway communication system (used, for example, by Foxboro, Honeywell) is one where nodes are organized into clusters, and clusters are linked into an integrated system. Each local cluster can combine controller modules, input/output modules, an operator CRT, and supervisory control module, for example. Communication within the cluster is fast, and intercluster communication is fast when clusters are organized properly. The cluster-based communication system is shown in Figure 2.8.

TOTAL DISTRIBUTED CONTROL

The major advantage of the distributed approach is increased reliability and security. If one module fails, the entire system does not fail. But the modular

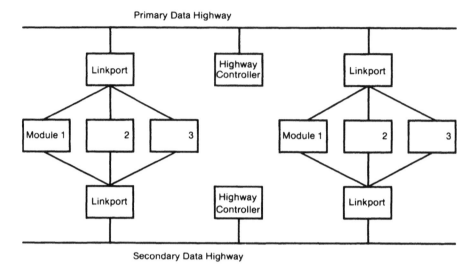

Figure 2.8. Cluster-based communication system.

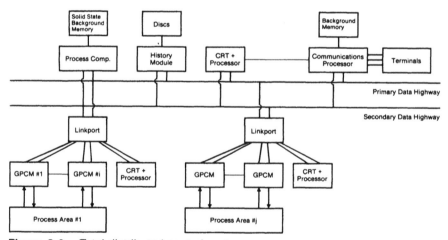

Figure 2.9. Total distributed control system.

approach results in simpler operating systems, which also gives more reliable operations and more predictable performance during disruptions. Application software can be tailored to the dedicated module function. The total distributed approach is shown in Figure 2.9. The plant is divided into *j* process areas. Each area is linked to a certain number of general purpose control modules (GPCM), which in its simplest form is an analog input module, for example, and in a more sophisticated form is a batch process control module with built-in backup memory that automatically takes over control upon failure of the primary memory.

Dual linkports link each process cluster to the primary and secondary data highway. Also connected to the data highway is a mainframe computer, a separate history and trend module, a CRT, and a communications processor. Local clusters can operate independently of each other, and critical GPCMs can be provided with backup GPCMs. Although error recovery is already built in at the hardware and module system software level, additional error recovery features can be built in at the application level, for example, by monitoring the module functions and switching to alternate control strategies in case of a failure.

CHAPTER 3

Computer System Software

Early computer systems were developed with an emphasis on hardware; the software was developed later to let the hardware function in the best possible way. In today's systems, however, hardware is designed to meet software constraints and demands.

System software can be classified into a number of levels:

- control software, which includes system control software and application control software. System control software is designed to let the system perform its specific functions, e.g., support program development, control program execution, and structure memory. This entire set of programs is called the operating system. Application control software is designed to control and monitor the process in the broadest sense.

- support software such as text editors, a link editor, language processors, e.g., Fortran, Assembler

- user-written software

A computer operating system will be examined more closely. The heart of the operating system is the real time executive. The real time executive structures and allocates memory, controls program execution, manages peripherals, detects a system malfunction, monitors power failures, provides memory and bulk storage, and so forth. A typical operating system structure is shown in Figure 3.1.

The real time executive monitors and controls seven subsystems that each control certain operations or support certain functions. The file manager subsystem allows the user to create, delete, and manipulate files. It can also copy and protect files and provide file security. The peripheral input/output subsystem manages the communication to and from all peripherals such as printers, typewriters, and system terminal. The console service subsystem services all consoles. It communicates with console memory, provides security, responds to key actions, and so forth. The process input/output subsystem provides an interface to the process. It manages analog and digital inputs and outputs. The asynchronous communication subsystem manages asynchronous communica-

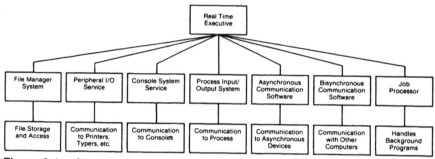

Figure 3.1.   An example of the structure of an operating system.

tion, and the binary synchronous communication subsystem manages high-speed communication with other computers. The job processor handles background programs, i.e., it executes and monitors them. Later in the chapter there will be further elaboration on this. The power-fail monitor monitors the power and, if a failure occurs, it takes appropriate action.

Application control software consists of software that is provided by the computer manufacturer and software developed by the user which has become part of the control package. This user-developed software is process-related, whereas application control software supplied by the computer manufacturer has a more general-purpose character.

User-developed application software can be written in different languages depending on the type of computer. For definition of the database, most computer manufacturers use some kind of "fill in the blanks" method. Many control and monitoring algorithms are standard on the system and all the user does is select parameters such that the algorithms fit his needs. For control of batch processes, most computer manufacturers have developed their own, often easy to use, language. For higher level continuous control tasks, the user will often have to write programs in Fortran or Pascal.

Support software is software that supports the user in his communication with the operating system. It provides valuable tools for efficient use of and communication with the computer. Most modern systems have text editors, which make file access and modification very easy. Some other utilities are Fortran, Pascal, and Assembler language processors; a link editor that links data, files, and programs to produce a module that is ready for execution; a console system editor for the creation of graphic, tabular, and bar graph displays; and a software maintenance system that documents all the software and gives the user the ability to maintain it. A software maintenance system is of ultimate importance since computer systems use hundreds of programs and files. Bad documentation will ultimately lead to inefficient use of the computer system.

REAL TIME PROCESSING

An on-line computer system communicates with the process on a real time basis via the process interface (analog-digital converter, digital-analog converter, digital I/O). In Figure 3.1 this communication is handled by a separate software subsystem although this is not always necessary. In order to communicate on a real time basis, the computer must have a real time clock, which is, in fact, system software that keeps track of the time based on incoming ticks (50 or 60 Hz). The clock enables the user to program events in real time.

A perfect illustration is a batch process. Suppose, for example, that a pump must be started, and two seconds later a valve must be opened. The real time application control program will give the command to start the pump, and two real time seconds later a command will be given to open the valve. The user will program a two-second wait in between the two control commands. During this waiting period, the application control program will leave memory to free it for other tasks. (It would be very inefficient to leave the control program in memory and to prevent other tasks from being processed. The control program is therefore swapped out of and into memory.) However, at the end of the waiting period, it should be in memory to process the "valve open" command. Real time processing has now become very important, since certain functions have to be performed within a certain time. The size of computer memory, memory structure, system and data highway loading and speed of communication with the process play an important role in this.

MULTITASK PROCESSING

As indicated in the previous section, a single user-single task system is very inefficient because a real time control program would require the utilization of all resources. If one wanted to develop software, the computer would have to be placed in the off-line mode where it is not controlling the process.

To get around this problem, many computer manufacturers designed a computer operating system that works with a foreground and background memory. The foreground memory is used to run critical control programs; the background memory can be used to perform other tasks, e.g., display building, text editing, or program development. Various mechanisms can be put in place to correctly handle the execution of programs. One could, for example, give foreground programs a priority and design a queue in which the programs are placed according to their priority. If a program with higher priority requests resource allocation, the program with lower priority can be suspended and be put in the queue in the appropriate location. Since the processing of background programs is not critical, they do not need a priority. They can be put in a queue, and a first-come-first-serve mechanism could be used to process the various tasks on a cycle basis. Once a program has been serviced for a certain time, processing is suspended and the program is placed at the end of the queue, after which processing of the next task is started.

These multitask processing systems require a relatively complex operating system. Much effort has been put into the software design to properly protect each task, to prevent the situation where two or more programs fight for memory allocation and the one program is preventing another from using memory, and so forth. It is evident that the user also needs to have a good understanding of the process and the computer in order to give all his control programs the right priority. Too many programs of the same priority might result in inefficient processing and delay in control tasks.

For all these reasons, there is a trend toward using distributed computer control systems where each processing system has a number of dedicated tasks. Because of the limited number of tasks, the operating system will be less complex with all associated advantages.

MEMORY PARTITIONING

Since there are a number of systems on the market with multitask processing facilities using a partitioned memory structure, this section will pay more attention to those systems.

Generally, the memory structure is as shown in Figure 3.2. The lowest area in memory is used by the operating system. The next highest area is dedicated

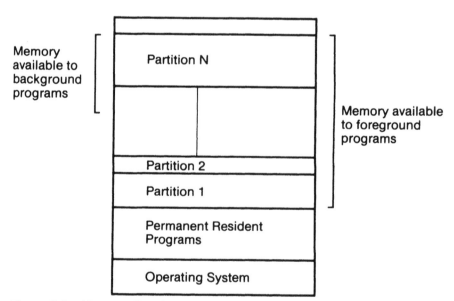

Figure 3.2.   Memory allocation.

to memory resident subroutines and programs, and then several partitions or sections of memory start.

Memory partitions are fixed in size, but can be different in size. As an example, partition 1 could be 4k, partition 2 could be 8k, and so forth. Programs with priority 1 can run in all partitions; programs with priority 2 can run in partition 2 and higher. Hence, the partition structure is determined by the total available memory and by the size of the programs. If we have a 14k program and want it to be able to run in at least two partitions, these two partitions need to be 14k or more. Background programs can run in the highest area in memory. The size of the background partition is usually fairly large. If all partitions in memory were in use and partition 2 and 5 were to be freed due to completion of the task, a new task could only be started if it fits into partition 2 or 5. However, if partition 2 and 5 were joined together, a much bigger task could be processed. To make this possible, tasks are shifted in memory after previous tasks are finished or suspended. This shifting or relocation can be done in two different ways: hardware relocation, in which each task is started at location zero and where a computer hardware register keeps track of the real location. Upon a relocation the register is used to update the new address at which a task is running. In other systems the relocation problem is solved via software. Since each reference toward a memory location is relative to the start of a program, the program can run at any location without a problem. There are a number of other techniques available for optimal use of computer memory, e.g., segmentation and virtual memory techniques.[9,10]

SOFTWARE DEVELOPMENT

Major Activities

Increasingly, software development is becoming the cost determining factor in a computer system. Some computer manufacturers already quote prices for system software packages that exceed the hardware cost. In the previous sections we went over some of the complex tasks the software has to perform. The design of this software is done by software teams, organized according to a certain structure. IBM, for example, introduced a hierarchical structure in a small team (see Figure 3.3). The chief programmer is a very experienced person and is responsible for the entire project. He is assisted by staff programmers, each one supervising and coordinating a team of computer programmers. A special function is that of librarian, who is responsible for the administrative work, the documentation, and so forth. In another organization, the matrix organization, people are selected according to function and project.[11]

In the process industries companies tend to purchase the system software

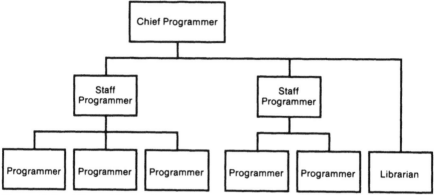

Figure 3.3.   Hierarchical software team structure.

and develop their own "application" software either with or without software support from the computer manufacturer. A software development team for this application software is usually small, and team structures are straightforward from project leader to programmers/engineers. For any medium- to large-size project, however, the team structure should be well established before the start of the project.

The functions that have to be performed in a particular project are independent of the size of the software project and the size of the software team. Before any specific computer application, a system must be selected based on specifications that have been defined by the process operations and control departments. Computer application engineers and project leader have a very important advisory role in preparing the proposal for system selection. In the final stages of system selection, a functional specification can be developed that defines what the different modules of the required software package will have to do.

Once the functional specification is complete, the development of the application software package can start. The major steps are the design, coding, and testing of the developed software (see Figure 3.4). When the software has been tested, it can be released for use and it must be "maintained." This is not maintenance in the true sense of the word, but implies making modifications due to mistakes in the functional specification, changes in process operation, and so forth. In other words, it is adapting software to the new environment, very often an ongoing task.

Very little has been published about implementing computer control systems in the process industry. Table 3.1 gives some data by Amrehn.[12] In this computer control project, about two-thirds of the time was spent in design of the plant control package. Table 3.2 is based on the author's own experience in designing and implementing a software package for a batch plant. About 43% of the total project time was spent by the team in designing displays, databases, and control programs and making changes in the software due to

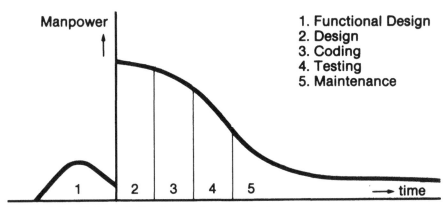

Figure 3.4.   Major activities in a software project.

Table 3.1.   Time Spent in Computer Project Activities

Activity	Total Project Time (%)
Computer specification	5.0
Process analysis	23.0
Programming	63.0
System installation	2.5
Off-Line testing	5.0
On-Line testing	1.5

Table 3.2.   Time Spent in Coordinating a Computer Control Project

Activity	Total Project Time (%)
Computer specification	5.0
Training project engineers	5.2
Software design	29.8
Database + display design	8.2
Software modifications	5.2
Project coordination + system installation	13.0
Field/program testing	10.3
Operator + plant engineers training	4.3
Working hardware and software problems	10.0
Improving system performance	9.0

Source: Author's experience.

changes in plant operation. Ten percent of the time was spent in solving hardware and software problems since the installed system had a very low service factor initially. Nine percent of the project time had to be credited to improving system performance. Load levels in communication with the instrument subsystems were balanced, and additional system software was developed and installed to increase system security. It is the author's experience that software design and testing depends very much on the nature of the process. In continuous plants, design, implementation, and testing is usually easier and less time consuming than in batch plants.

Software Design

The design of a software package should proceed according to a set of rules and guidelines that are established before the start of the project. The first step is a functional specification that defines what specific functions the software has to perform and also defines the functional elements that contribute to, or are part of, the final software package.[11,13] Similar or related functions should be allocated to modules in such a way that modularity and functional independence is enhanced. This will facilitate software changes that are easy to design and implement at a later date. The next step is to define interfaces between programs, and between program modules and the database. The third step is defining the modules. Once we have arrived at this step, each module can be defined and designed independent of other modules.

The functional specification of each module can be done using different techniques, e.g., flow diagrams (Figure 3.5). Boxes with different shapes represent the elementary processes; lines with arrows indicate the flow of infor-

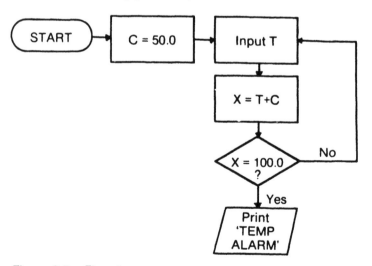

Figure 3.5.   Flow diagram.

mation. Flow diagrams can be used successfully if the modules have a limited size. In large modules a pseudo-code may give better readability. Normal English text is then used, such as "IF" . . . "THEN" . . . "ELSE" . . . and so forth. Pseudo-code has the disadvantage that it is difficult to get an overall picture of the structure; the advantage is that it is easy to read and is modifiable. It is also closely related to modern computer languages such as Pascal (Figure 3.6).

Yourdon[13] and Munson[14] give some guidelines in the design of software modules:

- Design according to a hierarchical structure if possible, e.g., top down.

- Limit module size.

- Each module should have one entry and one exit point, except in case of error exit.

- Each module should do its own housekeeping at the start of the module.

- Computational or input/output modules should not abort, but pass on an error code to the main program.

- Avoid difficult to maintain features, e.g., embedded constants.

- Access database by modules and limit the number of functions per module.

- Reduce interface/communication complexity by passing parameters directly between modules.

- Do not hard-code system level parameters into the modules.

- Use "go-to-less" or structured programming logic.

- Do not use implicit knowledge of one module in another.

Begin
$$C = 50.90$$
If X .neq. 100.0 then
    Input T
    $$X = T + C$$
Else print 'Temp Alarm'
End

Figure 3.6.   Pseudo-code.

- Minimize machine dependency.

- Forward jumps are allowed, but try to avoid backward jumps.

The major advantage of a modular design is that the software is portable or reusable with minimal effort in another system. Also, troubleshooting is easier since errors can be allocated to a module, and then the module can be examined closely.

The language in which the software should be written depends on the system. Many process computers use a special language for the application software that can be used in conjunction with Fortran or Pascal. Also, in these special application-oriented languages, we should strive for modular design, although some languages have restrictions in their use that might interfere with the design specification based on a modular structure.

It is not the intention to give a comprehensive survey on software design techniques that are described elsewhere,[13,15] but rather to familiarize the reader with the software design concepts that are of importance in each computer control project.

Software Maintenance

As explained in a previous section, software has to be maintained once it is designed and tested. This maintenance, however, is not maintenance in the true sense of the word since the software package will be able to perform exactly the same function over and over again.

However, process modification, for example, may bring with it the need to modify the software. Modular software design will certainly enable us to modify it without too much difficulty, provided the software was documented properly. Documenting software is something that has to be enforced during program development. The program designer does not have a direct need for writing comments in the listing, since he will understand the program in detail. However, the user will not be able to understand the program unless it is documented very well. When program changes have to be made comments are essential. Criteria for documenting source code listings are:[13,14,16]

- A computer program should contain a header of comments describing the following: program name, original design date and revisions, names of those who were involved in these activities, inputs and outputs, assumptions, methods used or algorithm definition, and error exit and recovery.

- Intermodule communication should be clearly defined.

- Each branching instruction should be annotated to show the structure of the logic.

- Variable names should be defined, and parameter ranges should be given.

- All program statements which have been added or modified should have an identification tag that associates the change with a documented change request.

Change request forms should be used with care since the amount of paper work easily gets out of hand. Minor changes that do not modify the structure of the program, but are necessary to perform a task better, should be made without a change request form but documented properly in the listing. For major changes that do change the structure of the program, a change request form should be filled out and filed. The source listing should contain comments describing the change and also why the change was made.

Some systems provide the possibility of the use of edits. These are small programs that only describe and code the change. During object module regeneration, these edits are then linked into the program where the modification is required. The advantage of this system is that the original program always remains the same. Edits can be added and deleted to create a final program that provides a different functionality. Here again the user should make sure that the number of edits does not get out of hand. It would also be advisable to incorporate minor changes directly into the original program. Adequate listings, however, are only half the problem. Listings will have to be supplemented with overview software system descriptions organized in terms of operational capabilities. This overview should specify the flow for each operational task, and the description should include the in-context references to the programs and database points and functions involved in processing each task. When a problem is encountered, the engineer should check the task listing which refers to the modules and data points involved and then check the listings of the specific programs for detailed information.

CHAPTER 4

Computer System Selection

It is difficult to give a clear-cut description of what is the best computer system for any given application. It will be dependent on whether the application is primarily batch or continuous in nature, the funds that are available, the required security, and so forth. Rather than listing minimal requirements for different categories of processes, a number of items that are of general interest to any user will be reviewed. The items are organized under the following headings:

- Hardware

- Software

- Instrumentation

- Project Organization

- Previous Experience

- Maintenance

- System Integrity

Each of the following sections will deal with these subjects in detail. System integrity will be dealt with when discussing hardware and software. However, it is worthwhile to review system response to hardware or software failures in somewhat more detail. A relative scale of importance could be assigned to the subjects mentioned above, and a particular system selected based on the outcome. One should bear in mind, though, that at the point where two systems differ, one could better take into account the cost and effort to bring the one system to the level of the other system and make a selection based on that outcome rather than to compare two different systems.

HARDWARE

There are a number of points in this category to which attention should be paid, especially the following:

- State of the art of the technology used in the system.

- Hardware Expandability. In other words, what is the maximum system configuration and how much of that will be used.

- System Modularity, Distribution and Redundancy. There are distributed systems on the market that use modules in which a fair number of the hardware circuit boards are interchangeable.

- Bulk Storage Reliability. Although new bulk storage devices have become available, they still have a lower availability than a computer.

- Peripheral Equipment. With increased processing speed, high-speed printers are an asset. Are they available as a standard product on the system you consider to purchase?

- Operator/Engineer Console and Console Human Engineering. Since the operator will spend a large amount of his time at the console, one should not underestimate the importance of appearance, color combination, and so forth. Special attention should be given to the console keyboard with respect to ease of use and layout.

- Capability of Interfacing with Instrumentation. Computer vendors usually provide interfaces with their own line of instrumentation. However, if there is already other instrumentation in the plant, or if one wishes to buy other instrumentation, it is good to check into this area. In too many cases, prospective users were satisfied with assurances that an RS232 interface could solve any problem in that area and found out later that they needed some specially designed hardware to solve the interface problems.

- Capability of Interfacing with Other Equipment. This can become important in communication to other computers. Computers do not necessarily have to be compatible with the system data highway. The same is true for other work stations.

- Limitations on Distance of Remote Stations. In many cases, there is a limitation on remote stations that could become important in interplant communications.

- Environmental Requirements. It would be advisable to pay attention to these requirements, especially in a plant environment. It is better to take precautions before the computer system arrives than when problems are identified in system operation due to high-level concentrations of a certain component.

- Power Requirements. One should be aware of the system power requirements and, if necessary, do a power line study on site.

- Failure Detection Capability. It is of ultimate importance for high system availability that failures are detected early and that the user is informed through meaningful messages. One should also check in which areas the system does perform error recovery.

- On-Line Failure Diagnostics Capability. In some systems, failure diagnostics can only be done when the system is taken off-line. In case of minor hardware problems, however, on-line diagnostic analysis would be a useful feature.

- Compatibility with Future Expansion. It is worthwhile mentioning that two years after installation any process control system is usually too small. The questions should be asked if it is possible to expand memory without major problems, to increase the number of operator stations, and so forth.

SOFTWARE

Since the available software determines the power of the system, the rating on the relative scale of importance would be higher than for the hardware. Software design and testing is usually very time consuming, and consequently a good standard software package is of extreme importance.

With special reference to a process control computer, the following areas should be considered in computer selection:

- Continuous Control Functionality. It is important to check if it is easy to implement continuous control via predefined algorithms, if direct digital control can be changed into setpoint control and reverse, and so forth.

- Subsystem Communication Capability and Compatibility. Especially in a distributed system, one should pay attention to interunit communication. Can one subsystem initiate a task in another? Can processors in a network access each other's peripherals? Can processors share databases on bulk (global database)? Can remote processors detect and act on the failure of another processor? Does the local operating system ensure that network requests do not swamp the processor?

- Scan and Alarm Capability. How many process points can the system scan per second? Is it possible to take points temporarily out of service? Is there a facility in the system that informs the operator which points or tags are out of service?

- Higher Level Supervisory Control Language. It is important that supervisory programs can be written in a higher level language, e.g., Fortran or Pascal. Check whether it is possible to use subroutines written in a higher level language in specific application-oriented languages. The latter languages are usually very easy to use. However, complex mathematical calculations are often cumbersome in these languages, therefore it would be easier to write them in Fortran or Pascal. It would also be useful if supervisory programs could access the historic database.

- System Software and Operation Security. Since the computer is communicating to instruments in the field, the system should be protected against erroneous outputs. It is also relevant to check what kind of action should be taken if discrepancies are found between a database in memory, for example, and on bulk storage. Do control programs run in a protected area that cannot be overwritten by other programs? There are numerous other questions into which we could check. (See Chapter 7 for more detail.)

- Support Software Capability and Efficiency. One should check into support software provided with the system: text editing facilities, file building facilities, and so forth.

- On-Line Program Generation and Modification Capability. Is it possible to modify control programs and databases on-line?

- Operating System. Compare the structure of the operating systems of the computers in which you are interested.

- Program Scheduling Capability. What is the maximum number of programs that can be scheduled at a certain frequency and what will be the impact on system loading?

- Historic Database Functionality. What is the trending capability of the system? How easy is it to use? Can it interface with control and user programs? Are standard reports and displays available?

- Console Display Flexibility and Display Building Capability. Are graphic displays standard on the system? What other types are there? How easily can displays be built and modified?

- Tag Building Flexibility. On a process and instrument diagram temperature transmitter, 1506 would be indicated by TI1506. Does the system allow this number of characters for tags? Some older systems only allow three or five characters per tag. A modern system should allow at least six but preferably a larger number of characters per tag. Some computer systems restrict the number of characters that can be used in each separate location of the tag name. For example, the user has to set up eight- or sixteen-character symbol lists for each character in the tag name. The higher the number of characters in this list, the more flexibility you will have in selecting tag names.

- Batch Control Capability. If your application is a batch application, a batch software package should be available for the system. Check how easy it is to implement your own application programs and how easily they can be modified.

- Debug Flexibility. Attention should be paid to this area since it can be very frustrating to try to correct errors in programs when hardly any debugging tools are available or no meaningful error messages are generated.

- System Restart Flexibility. It is possible to specify how the system should come up after a crash or a power failure? Is it possible to distinguish between long- and short-term power failures?

- Automatic Switching on Peripheral Failure. Can one peripheral back up another one, e.g., can a system terminal take over the duties of a printer when the printer fails?

- Instrumentation Interface Diagnostics. Can diagnostics be used to determine problems in an instrument subsystem?

- Documentation/Diagnostics in Case of System Crash. Although a system failure will not be an everyday occurrence, it is important that the system produce a memory dump in case of a failure and that adequate material is available to analyze the dump in order to find the problem.

- Effective System Loading and Loading Monitoring Tools. In a real time environment it is of ultimate importance that control programs run on time. As the system loading increases, the potential for delays in the execution of programs will increase. Does the system have a monitoring and alarm tool to detect this, and does the system automatically distribute the execution of programs in such a way that the loading is minimized and peak loading avoided?

- Recipe Facilities and Flexibility. Recipes play a role in a batch environment where one could have recipes for different products. Make sure that various recipe formats are available and that the recipe size can be specified.

- Software Documentation. If a system is maintained by site personnel, particular attention should be paid to documentation. It will be useful to have the following documentation available:
  1. Source code and executable object code (program listings)
  2. System software documentation
  3. User manuals
  4. System hardware documentation

5. System installation drawings, electrical drawings, drawings for maintenance personnel
6. Settings of selectable operating modes in hardware

INSTRUMENTATION

Many items discussed in the previous sections could be repeated for the instrument subsystems, e.g., console and console human engineering, operating modes, computer interface capabilities, trend capabilities, configuration flexibility and functionality, expandability, and alarming. A few items should be mentioned more explicitly.

• Power Supplies. Is dual power supply from different sources available and is it distributed? Make sure, for example, that a common power supply is not used for several groups of digital inputs but that each group has its own power supply.

• Intrinsic Safety. Especially in a plant environment, intrinsically safe equipment should be looked for.

• Instrument System Backup. One hears often about control computer backup, but insufficient attention is paid to instrument backup. Is it possible to back up critical process values in an easy way?

PROJECT ORGANIZATION

Especially on a large project the vendor organization will involve many people and a number of factors should be considered:

• Capability to Provide Software/Hardware Assistance. Does the vendor have qualified and sufficient resources to provide assistance in these areas if required? During and after computer system installation these resources will be required until the user has built up a certain level of confidence in operating the computer system.

• Training Courses. Does the vendor provide training courses for personnel who have to work with the system?

• Schedule. Does the vendor have an organization that is capable of stewarding toward the target dates that are set? Large computer/instrument systems are not available off the shelf but are manufactured after they are ordered. Delays in delivery date will usually lead to additional expenses.

- Vendor Reputation. It is obvious that only vendors who have a good reputation and the desired resources are considered.

PREVIOUS EXPERIENCE

It is worthwhile to check with other users to find out what their experiences have been with the system you want to purchase and with the company that manufactures the computer/instrument system. Try to find out what problems they experienced and improvements they have identified. Compile a list of problem areas and determine which of those are serious enough to bring up in your discussions with the vendor. If at all possible, get problems solved as part of the standard hardware and software. Special custom hardware and software may solve certain problems initially, however, there is always the problem of maintainability. Within a number of years there will probably be no one who knows anything about your special custom, and in case of problems, it would be very time consuming and expensive to get the problems fixed.

MAINTENANCE

Hardware and software maintenance must be considered before the purchase is made. If the system is maintained by your own personnel, computer and instrument spare parts should be kept on site. However, it is impossible to keep all spare parts in stock, hence a check should be made where spare parts are available and in what quantity. The availability of spares plays an important role in maintaining a high system service factor. Even if you choose to have a maintenance contract with the vendor, it is advisable that some of the most critical spare parts be kept on site. Negotiate a price for the maintenance contract and come to an agreement with respect to response time, price escalation, equipment to be serviced, price for spare parts and so forth. Get written assurance from the vendor that spare parts are available for at least the next ten years. Finally, determine what is necessary for preventive maintenance to be done on the system and agree what utilities you have to provide.

It may also be worthwhile to look into system software support. Is it required and how much will it cost? Determine for how long you will need support and how long the vendor is willing to give you support. The answers to some of these questions may help you to decide to train your own personnel rather than using the computer manufacturer's services.

SYSTEM INTEGRITY

Integrity has to do with system response to detected hardware or software problems. Error recovery is just as important as error detection. Some questions a prospective user should ask are:

- What happens on a short- or long-term power failure? What is the transition time between the two, and how does the system respond in case backup power supply is available?

- What happens if the bulk storage device holding the backup database and control programs has a failure?

- What happens to all historic data during a computer failure?

- How does the system detect corruption of the process database and control programs? What does happen in this case?

- What hardware integrity checks are built in? What happens if a check fails?

- What software integrity checks are there?

- What happens if the central processor stalls? How is control passed to a backup or remote processor?

- Is the system protected against outputting erroneous values to field instrumentation?

Compare the answers to these questions with your expectations and requests. Once you have bought the system, it will be extremely difficult to fix problems related to integrity and security.

OTHER CONSIDERATIONS

Two things that remain unmentioned so far are system staging and testing. Especially in case of a large computer/instrument system, there may be the need on your part to test the system before it is shipped to you. Define what you want to test and for how long, and what your understanding is of the successful test, i.e., when you can accept shipment of the system. Test the integrated system again for a longer period after it is installed on site and aim, for example, for a system service factor of at least 99% over a period of two months. Define exactly what will be considered as system downtime, a critical failure, and so forth. Some thinking ahead will pay off and produce a satisfactory system. It is important to talk these things over with the computer manufacturer and have a common understanding of the responsibilities of each party involved. When you have a smaller system, one shorter test on site may be sufficient. It is difficult to give a general guideline in this area. You as a computer user, however, have to make sure that you have a system that performs the duties for which it was designed and delivered.

CHAPTER 5

Reliability and Security

Modern control computers allow the implementation of control strategies impossible several years ago. With this increase in complexity, the consequences of failures of the system have also become more severe. A failure is usually limited to one controller when using analog instrumentation, and process operators now have to put this one loop into manual control.

The first process computers were used for direct digital control (DDC). Failure of the computer had very serious consequences for process operation. Usually the total process was uncontrolled. To avoid this problem the control philosophy changed. Conventional analog instrumentation was used for control, and the computer was used for the adjustment of setpoints (SSC = setpoint supervisory control). If the computer failed, the process was still being controlled, and the operators could adjust setpoints if required. This approach is still applied in control systems today. At a later stage DDC was applied again, this time with analog or digital backup. The computer controls the process and, in case of a failure, control is reverted to analog equipment or a backup computer. Figure 5.1 gives the latest set-up: distributed direct digital control. Systems based on mini- or microcomputers each control part of the process and are supervised by a computer which could function in the role of setpoint supervisor. This set-up is chosen not only in continuous control, but also in batch control.

If critical process variables are present in a part of a process, that part can be provided with backup control, which will now also have a distributed character. Due to continued technological developments, computers will become increasingly reliable. Whenever reliability is analyzed, it is important to include the data channel or channels in the analysis.

The next section will give a detailed analysis of the concept of reliability. In that context the definition of the "failure of a computer system" will be dealt with also. The definition of failure has lead to many discussions in the past. Between a total failure and a number of small faults is a large spectrum of possibilities. The important question is then, When has the computer system failed? or, When do we speak of a "system failure"? It is possible, for example, to create a quantitative limit: a system fault is a failure when at least 30% of the controls have become inactive. In most reliability analyses one distin-

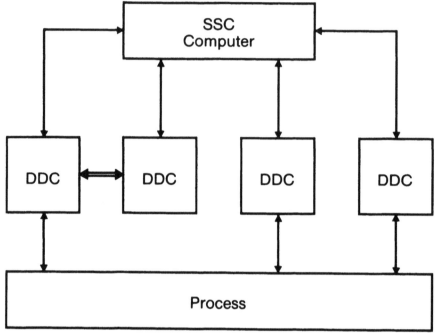

Figure 5.1.   DDC process control with digital back-up.

guishes only between failed or not failed. It is obvious that the aforementioned problem should urge us to carefully use data from a reliability analysis.

DEPENDABILITY

A system is usually evaluated in terms of its dependability. Dependability is more than reliability or availability – it also includes security and integrity. Availability is a measure of how often the system is available for performing the functions it is supposed to do. Security has to deal with mechanisms that guard against unauthorized system use or inadvertent misuse. Diagnostic and security actions have to be taken when malfunctions are detected to alert the operator and protect the system (or maintain functional system integrity). System security comprises hardware and software components. One example in this area is error prevention through keylocking functions.

Integrity is the confidence that the system is performing correctly. To check this the system will continuously check its operating system for malfunction, e.g., by activating a watch dog timer. The timer should be reset by the system software at regular intervals. If the watch dog timer times out, the computer processor will halt.

Each of the cornerstones of dependability will be considered in more detail.

RELIABILITY ANALYSIS

In reliability analysis[17-19] the Markov method[17] is often used. The starting point is that an element can be in only one of two discrete states: good or bad, available or not available. It is further assumed that during a time interval Δt, only one occurrence will happen and that all occurrences are independent. The probability that a failure will occur, the so-called failure rate, is assumed to be constant and equal to μ. In general, this failure rate will not be constant. Immediately after a computer installation, a large number of failures will occur due to bad components, bad connections, and so forth. Also, the software may have several errors in it. (Compare Table 5.1 which gives the author's experience with two identical systems.) After some time, there is usually a period during which μ is fairly constant, but, as the computer gets older, the number of failures will increase due to the aging of components. The failure rate μ, therefore, has the shape as shown in Figure 5.2. In the reliability analysis, the restriction is usually made to the time period in which μ is constant. In a Markov flow diagram, all possible states (good and failed) are presented together with the failure and repair rate. Using this flow diagram,

Table 5.1.   Comparison of Availability of Two Identical Computer Systems Installed in Two Different Plants

	Computer System A		Computer System B		
Year	Downtime (DT) (hr)	Availability (%)	Downtime (DT) (hr)	Availability (%)	MTBF (hr)
1981	619.0	92.9			
1982	445.0	94.9			
1983	105.6	98.8	415.5	95.3	593
1984	8.7	99.9	13.4	99.8	2184

	Failure Rate System A (μ yr^{-1})	Failure Rate System B (μ yr^{-1})
Core memory	0.75	0.50
Asynchronous nest communication	0.50	0.50
CRT	0.50	—
CPU	0.25	—
Disk	0.24	0.66
Substation	0.25	—
Overheating	0.25	—
Floating point hardware	0.25	0.50
Software	0.75	1.00

Source: Author's experience

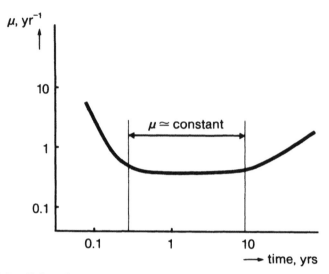

Figure 5.2. Failure frequency as a function of time (bathtub curve).

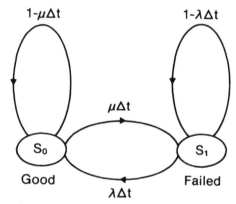

Figure 5.3. System flow diagram.

we can derive a state table from which the differential equation that describes this process can be derived.

SINGLE SYSTEM

The flow diagram for a single system is shown in Figure 5.3. The failure rate (number of failures per year) is μ, and the repair rate is λ (= total time in use/ average repair time). State S_0 indicates that the system functions well; state S_1 indicates a failure. The probability that the situation changes from S_0 to S_1 during the time interval Δt is equal to $\mu \Delta t$. The probability that the situation changes from S_1 to S_0 in the same time interval is $\lambda \Delta t$.

$P_0(t)$ is the probability that the system functions in time t; $P_1(t)$ is the probability that it has failed in time t.

The probability that the system functions can be determined with the aid of Table 5.2 and is given by the differential equation:

$$P_0(t + \Delta t) = (1 - \mu\Delta t)P_0(t) + \lambda\Delta t\ P_1(t) \qquad (5.1)$$

or in the limit case:

$$\dot{P}_0 = -\mu P_0 + \lambda P_1 \qquad (5.2)$$

in which the dotted variable denotes the derivative of the variable with respect to time.

In a similar way the probability can be determined that the system has failed:

$$P_1(t + \Delta t) = \mu\Delta t\ P_0(t) + (1 - \lambda\Delta t)\ P_1(t) \qquad (5.3)$$

or

$$\dot{P}_1 = \mu P_0 - \lambda P_1 \qquad (5.4)$$

Summation of equations (5.2) and (5.4) gives:

$$\dot{P}_0 + \dot{P}_1 = 0 \qquad (5.5)$$

or

$$P_0 + P_1 = constant = 1$$

Table 5.2. State Table for One System

	Final State at Time t + Δt	
Initial State at Time t	S_0	S_1
S_0	$1-\mu\Delta t$	$\mu\Delta t$
S_1	$\lambda\Delta t$	$1-\lambda\Delta t$

which will not surprise us. The solution of equations (5.2) and (5.4) is:

$$P_0(t) = \frac{\lambda}{\mu + \lambda} [P_0(0) + P_1(0)] +$$

$$\frac{\lambda}{\mu + \lambda} e^{-(\mu + \lambda)t} [\mu P_0(0) - \lambda P_1(0)] \qquad (5.6)$$

and

$$P_1(t) = \frac{\mu}{\mu + \lambda} [P_0(0) + P_1(0)] +$$

$$\frac{1}{\mu + \lambda} e^{-(\mu + \lambda)t} [\mu P_0(0) + \lambda P_1(0)] \qquad (5.7)$$

If the system initially functions, then:

$$P_0(0) = 1 \quad \text{and} \quad P_1(0) = 0 \qquad (5.8)$$

which reduces equations (5.6) and (5.7) to:

$$P_0(t) = \frac{\lambda}{\mu + \lambda} + \frac{\mu}{\mu + \lambda} e^{-(\mu + \lambda)t} \qquad (5.9)$$

and

$$P_1(t) = \frac{\mu}{\mu + \lambda} - \frac{\mu}{\mu + \lambda} e^{-(\mu + \lambda)t} \qquad (5.10)$$

The static values (as t approaches infinity) give the average availability A:

$$A = P_0(\infty) = \frac{\lambda}{\mu + \lambda} \qquad (5.11)$$

Equations (5.2) and (5.4) can be written in a simple form in matrix notation:

$$\begin{bmatrix} \dot{P}_0 \\ \dot{P}_1 \end{bmatrix} = \begin{bmatrix} -\mu & \lambda \\ \mu & -\lambda \end{bmatrix} \begin{bmatrix} P_0 \\ P_1 \end{bmatrix} \qquad (5.12)$$

The mean time to failure (MTTF) can be found by substituting $\lambda = 0$ in equation (5.9):

$$P_0(t) = 0 = e^{-\mu t} \qquad (5.13)$$

$$MTTF = \frac{\int_0^\infty t \, dP_1(t)}{\int_0^\infty dP_1(t)} = \int_0^\infty t e^{-\mu t} \, dt = \frac{1}{\mu} \qquad (5.14)$$

In a similar way, we find for the mean repair time (MRT or MTTR):

$$MRT = \frac{1}{\lambda} \qquad (5.15)$$

The mean time between failures (MTBF) follows from:

$$MTBF = MTTF + MRT \qquad (5.16)$$

Substitution of equations (5.14) to (5.16) into (5.11) results in:

$$A = \frac{MTTF}{MTBF} = \frac{MTBF - MRT}{MTBF} = \frac{MTBF}{MTBF + MRT} \qquad (5.17)$$

In order to decrease the mean time to repair and to provide increased safety, built-in diagnostics and integrity tests are becoming increasingly important.

REDUNDANCY

To avoid complications, it is assumed that two identical parallel systems are being dealt with, each with a failure rate μ and repair rate λ. The flow diagram for all possible situations that can occur is given in Figure 5.4. These situations are:

S_{00} both systems are good
S_{10} the first system has failed
S_{01} the second system has failed
S_{11} both systems have failed

When the first system fails, the second one is used. The state table is given in Table 5.3. From this table the following equations can be derived:

$$P_{00}(t + \Delta t) = (1 - 2\mu\Delta t) P_{00}(t) + \lambda\Delta t P_{10}(t) + \lambda\Delta t P_{01}(t) \qquad (5.18)$$

In the limit case:

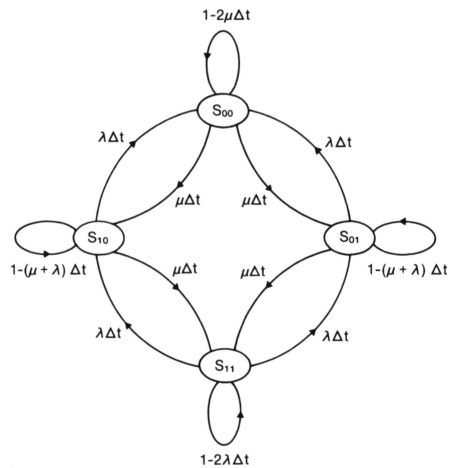

Figure 5.4.   Flow diagram for two parallel identical systems.

Table 5.3.   State Table for Two Parallel Systems

	Final State at Time t + Δt			
	S_{00}	S_{10}	S_{01}	S_{11}
Initial State at Time t				
S_{00}	$1-2\mu\Delta t$	$\mu\Delta t$	$\mu\Delta t$	0
S_{10}	$\lambda\Delta t$	$1-(\mu+\lambda)\Delta t$	0	$\mu\Delta t$
S_{01}	$\lambda\Delta t$	0	$1-(\mu+\lambda)\Delta t$	$\mu\Delta t$
S_{11}	0	$\lambda\Delta t$	λt	$1-2\lambda\Delta t$

$$\dot{P}_{00} = -2\mu P_{00} + \lambda P_{10} + \lambda P_{01} \tag{5.19}$$

Similar equations can be derived for P_{10}, P_{01}, and P_{11}:

$$\dot{P}_{10} = \mu P_{00} - (\mu + \lambda)P_{10} + \lambda P_{11} \tag{5.20}$$

$$\dot{P}_{01} = \mu P_{00} - (\mu + \lambda)P_{01} + \lambda P_{11} \tag{5.21}$$

$$\dot{P}_{11} = \mu P_{10} + \mu P_{01} - 2 \lambda P_{11} \tag{5.22}$$

The average availability can be determined by solving the static versions of equations (5.19) to (5.22) ($P_{ij} = 0$, $i = 0$, 1 and $j = 0,1$) under the condition that the sum of the average possibilities is equal to one:

$$P_{00} + P_{10} + P_{01} + P_{11} = 1 \tag{5.23}$$

The result for the average probability of a total system failure P_{11} is:

$$P_{11}(t \rightarrow \infty) = \frac{\mu^2/\lambda^2}{(1 + \frac{\mu}{\lambda})^2} \tag{5.24}$$

hence, the average probability that the system functions well, or the average availability, is:

$$A = 1 - P_{11} = \frac{1 + \frac{2\mu}{\lambda}}{(1 + \frac{\mu}{\lambda})^2} \tag{5.25}$$

It should be noted that it is assumed that the switching over from one computer to another is ideal. In other words, it is assumed that the switching failure rate is only a fraction of the failure rate of the system. Bad switching could even result in a redundant system that is less reliable than a single system. If the failure frequency is of the same order of magnitude as the system failure frequency, the equations will be much more complex.

As an illustration, consider a computer system that is in use for 8400 hours per year. On the average there are three failures per year. The average repair time is 36 hours, which includes travel time of the technician. According to the definition, $\lambda = 233.33$ yr^{-1} and $\mu = 3$ yr^{-1}. Hence, for a single computer system:

$$A = \frac{\lambda}{\mu + \lambda} = \frac{233.33}{236.33} = 0.987$$

For two identical systems under similar conditions:

$$A = \frac{1 + 2\frac{\mu}{\lambda}}{(1 + \frac{\mu}{\lambda})^2} = \frac{1.0257}{1.0259} = 0.9998$$

From these results it is clear that a single system is unavailable 1.3% of the time; a redundant system is unavailable only 0.02% of the time.

INTEGRITY

With computer systems becoming increasingly complex, there is a great need for integrity checks, which are performed in order to analyze problems faster and reduce the time to repair. The purpose of these checks is to determine whether there is a hardware or software error in the system. Usually error counters are maintained which display, for example, parity errors, security disconnect errors, or time out errors. The system will usually try to reset these counters and alert the operator if the reset is unsuccessful.

Most systems are able to correct single-bit memory errors without affecting processor operation. Some other useful integrity checks on modern systems are: power failure detection and recovery, regular disk read and write checks, and a check on the correct computation of control algorithms. As integrity checks become more comprehensive, the actual availability of the system will increase.[20]

LITERATURE DATA

Amrehn[21] gives data for a number of process computers of six different computer manufacturers. The systems were used for a variety of chemical processes. Whereas one system had an average availability of 92.47%, for all other systems the availability varied between 99.1% and 99.9%. The MTBF varied between 12.3 days and 257.3 days and the MRT between 1.07 and 31.2 hours.

The mean repair time will largely depend on the way a system is maintained. Consider a system with four failures per year. On one site, spare parts are available, and the system is maintained by the company's own technicians. This leads to an average repair time of three hours. On another site, no spare parts are available, and the system is maintained by the computer manufacturer's representative. In this case, the average response time is 8 hours, but the actual repair time is 20 hours because spare parts are sometimes unavailable and have to be shipped in by courier. In the first case, the availability is

99.85%, in the second case 98.68%. It is of ultimate importance that these factors are considered during the purchase of a computer system and that the cost of training of at least two technicians and keeping a spare parts inventory are compared with the cost of lost production and buying a maintenance contract.

Amrehn[12,21] gives data for a redundant computer system. The first computer experienced seven outages with a total downtime of 75 hours; the second computer experienced five outages with a total downtime of 38 hours. The entire computer system experienced two outages with a total downtime of 31 hours, showing that the availability of the total system (99.63%) is hardly better than that of the second computer (A = 99.55%). It could have been that there were common mode failures in this case. When a redundant computer system uses, for example, a common disk, then on a disk failure the total system will fail. Therefore, in a redundant system, all elements have to be able to function independently.

Other data is given by Escher and Weber[22] and Stübler.[23] Escher and Weber[22] give the availability of two types of Foxboro computers and a Kent computer. Results are summarized in Table 5.4. Stübler[23] gives the availability of 15 computers during a 3.5-year time frame. During this period, a total of 222 disturbances were registered of which 143 lead to a failure. The average availability turned out to be 98.7% which is low compared to Amrehn's data.[21] One explanation for this low availability could be that one of the 15 computers contributed to more than 40% of the failures. Stübler[23] summarizes faults that lead to failures in hardware and software. Results are shown in Table 5.5. Application programming errors contribute to about 25% of the total number of problems, and about two-thirds of all faults lead to a real system failure.

Apart from the possibility of increasing the availability by applying a redundant system, Plogert and Schuler[24] and Frey[25] give some other examples of increasing availability, e.g., by applying a two-out-of-three system. (See also Roffel and Rijnsdorp.[26]) In a two-out-of-three system, three computers work in parallel, and the system is active as long as two computers produce the same result. Plogert and Schuler[24] and Frey[25] discuss these systems extensively, and it turns out that a two-out-of-three system is not more reliable than a redundant (one-out-of-two) system under all circumstances. The installation of three

Table 5.4. Availability of Process Computers

Computer	Hours in Operation	Failures	MRT	Availability
Foxboro PCP88	35,000	27	1.1	99.92
Fox 2/30	16,400	76	1.5	99.29
Fox 2/30	3,600	1	0.5	99.99
Kent K75	22,600	5	5.8	99.87

Source: Escher and Weber.[22]

Table 5.5.   Faults that Lead to Failures in Computer Systems

	Failures (%)	Faults (%)
Electronic parts	7.9	12.9
Electromechanical parts	1.9	12.9
Contact problems	1.9	4.9
Voltage supply and temp control	5.9	6.9
User error	3.0	3.4
Operating system	6.9	7.7
Application software	15.4	22.3
Maintenance	4.1	9.5
Unknown	17.4	19.5
Total	64.4	100.0

Source: Stübler.[23]

parallel systems is still very costly and therefore will not be applied yet. A better approach is the distributed approach as shown in Figure 5.1, where DDC micro- or minicomputers control the process, and for critical areas backup is provided.

SECURITY

In the section on integrity, the detection of errors, either in hardware or software, was discussed.

Another aspect of computer systems is the protection of software and hardware. The protection against misuse of the system and misuse of data is called security.[20] In software the protection is usually realized with the aid of key words and protection codes. The key words enable a user to get access to the system, whereas the protection code is used to protect existing data against reading the data or writing into or over the data. Another kind of protection of the software against errors is called safety. Some methods will be described.

It is important that the many tasks that are processed in parallel in a real time system do not disturb each other or interact in an undesirable way. Methods to prevent or restrict these errors are sometimes carried out in software. These solutions, however, are not always foolproof. Therefore, hardware methods are also possible. One hardware method is so-called memory protection. A certain protection register makes sure that a user or program can only address a fenced area in memory. When reference is made to an address outside the fenced area, an error detection is given after which the operating system aborts the task.

Apart from these memory protection facilities, memories can also be used with write protection. By dividing a program into an operation code area and data area, it is possible to protect the entire memory, with the exception of its own data area, against write or modification commands. One task will then never be able to cause errors in the software of another task or even overwrite its own operation code. With special instructions or switches, write command for disks can be prohibited (write protection). A possibility for detecting memory errors is the use of parity bits. With the information, one or more bits are set. The parity bits are set during a write command in such a way that the number of ones in a unit of information (e.g., a word) is either odd or even. During the reading command, a test is made whether the parity is consistent. In this method error correction is not possible, because the errors are hardware errors.

Another kind of protection is important in data communication between two computers. The communication is only established when the computers have been introduced to each other with the aid of key codes (hand shaking). The transmitted code is checked for correctness with the parity check. To make this even more secure, an extra code is transmitted (redundant code). The sender produces this code with the aid of the transmitted data according to a certain algorithm, and the receiver checks the code according to the same algorithm. By selecting a suitable algorithm, it is possible to introduce error detection and error recovery.

CHAPTER 6

Batch and Continuous Process Control

Automatic control of continuous processes dates back to the early 1950s. Over the past 20 years computer systems have been continually adapted to process control, however, with an emphasis on continuous processes. In the late 1960s the programmable controller (PLC) was developed in response to the automotive industry's requirements. In the early 1970s PLCs were first applied in control of batch processes in the chemical industry. At that time (late 1960s/early 1970s), a limited number of computer systems that offered a batch control software package was available.

A major breakthrough came in the early 1980s when computer manufacturers developed batch control systems for mini- and microcomputer systems. Until recently, there were a number of differences between a PLC and a computer system:

- A PLC has a simple operating system, and only limited security checking is done. To bring the security checking of both hardware and software functions within a PLC at the level of a computer would at least require hardware modifications and major (expensive) software modifications.

- A PLC has a simple and specific application language in the form of a ladder diagram or Boolean algebra. This implies a limited instruction set.

- The operating system of a PLC is not accessible to a user. Higher level programming languages cannot be used.

The scope of batch computer applications is usually different from the scope of a PLC application. A computer can be used for trending and documentation of process variables, report generation, more sophisticated control through supervisory (Fortran or Pascal) programs, and so forth, but until recently these functions were not available in PLCs. The latest development is that computer manufacturers develop interfaces between PLCs and computer systems. PLCs are then used for sequences which are repeated continually; the computer monitors and controls the PLCs by initiating sequences and performs the data collection and information processing. The advantage of this approach is that the computer is freed from several tasks and thus can be used for other tasks. A PLC is also an order of magnitude less expensive than a computer system. One should keep in mind, however, that the security and integrity checking of a PLC is limited, although continuous improvements are being made in this area.

Another recent development is that PLC manufacturers designed interfaces to personal computers providing color graphics and documentation facilities. Modules are now available to interface PLCs to mass storage devices.

Since 1984 there have been PLC-based systems on the market which perform all the functions of a conventional programmable controller, with additional capabilities previously found only in mainframe process control computers. They are characterized by a design architecture that combines the features of a programmable controller (ruggedness, high-speed scan ability, low-cost I/O, ease of programming) with the most desired features of a process control system (communications capability, higher level programming languages, large memories for data storage).

SOME SPECIFIC PROBLEMS IN BATCH CONTROL

Batch processes have been defined in many different ways (see, e.g., Mehta[27] and Roffel and Rijnsdorp[28]). It will be sufficient here to define a batch process as a process where the operation is time dependent and repeatable.

An example could be a chemical reactor that is being filled with ingredients and heated, after which a reaction takes place. Upon completion of the reaction, the reactor contents are dumped, the reactor is cleaned, and the same sequence of events is repeated. The analogy between a batch process and the start-up of a continuous process is that in both cases the operation is time dependent; however, in a continuous process, start-up occurs only once.

Not in all batch systems will there be conversion of components. Two main categories of processes can therefore be distinguished: batch processes in which components react and processes without reaction. When components react, there is a chemical reactor which is filled either automatically or manually. The conversion of material is initiated by catalyst and/or heat. Upon reaction takeoff, continued heating may be necessary for endothermic reactions, or cooling in case of exothermic reactions. The reaction time is usually large (6–10 hours) compared to the charging time (e.g., 15–60 minutes). In order to produce the required quantity of product, more than one reactor is often used, whereas the charging subsystem (the system which charges all the components) will be shared by the reactors due to the relatively short charging time.

A scheduling problem may exist when the number of reactors becomes large. Assume, for example, that the turn-around time for a reactor is eight hours, i.e., every eight hours a reactor is charged. Charging takes one hour, and there are eight reactors. If all reactors were perfectly scheduled one hour apart, the charging unit would just be able to manage all the charges. When, under the same conditions, some reactors have a shorter turn-around time, a scheduling problem would exist: Which reactor is charged, for example, when two make a request at the same time? If the reactors are identical, a first-come first-served queuing mechanism could be used to serve the reactors. When the reactors are different in size and when the reaction time varies per reactor, for example, due to limitations in heating or cooling, an optimization problem may exist in which product quantity can be optimized.

When the reaction is complete, the reactor contents are usually dumped into one or more intermediate storage tanks. Here the same problem can exist as in the case of charging: one or more reactors may want to dump at the same time. However, there may be room for the contents of only one reactor. Figure 6.1 shows a sequential "process" with common charge and dump facilities: a cup of coffee is prepared by adding cream to the coffee. Only one cup can be charged at a time. Since more than one cup can be waiting for coffee or cream, mutual exclusion is applied: when the resource is available, a flag is set; when the resource is used, the flag is reset. There is a strong analogy with software principles; for example, two computer programs may be waiting to use file

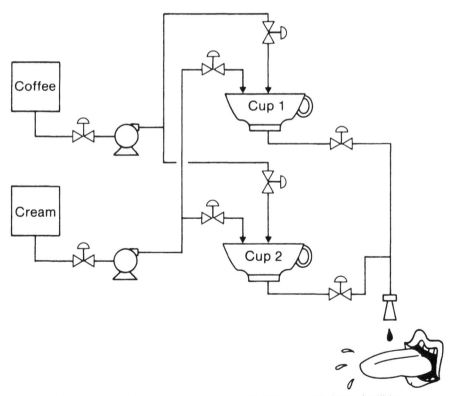

Figure 6.1. Sequential process with common charge and dump facilities.

space or a peripheral. In software engineering, many queuing models exist, and solutions are well known to resolve the contention for resources. However, these solutions and principles are little used in batch process control.

Consider a second type of batch process in which no ingredients react. Processing times are usually short. A number of specific examples will be given.

The first example is a batch drying process in which a batch of material is treated with hot air until a certain moisture level is obtained. If there is no limitation in the hot air supply, then even for parallel units the queuing problem does not exist. A similar example is a steel plate which has to be heated to a certain temperature before it can be treated further. Another example is the preparation or batch mixing of components. Two or more components may be added to a tank, mixed, and transferred for further processing. Another example is a resin transfer facility where resin is transported from intermediate storage tanks to final product storage silos. In batch systems with short to intermediate processing times, sequence control is usually straightforward; when processing times (as in chemical reactors) become large, parallel processing units may become necessary to achieve the required production, and interaction between the parallel process units and shared process units needs to be well defined in order to avoid problems.

One approach to control of a batch plant, as shown in Figure 6.2, was as

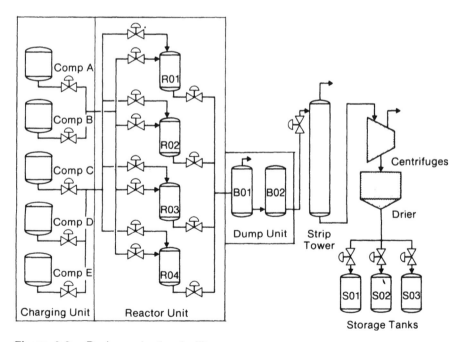

Figure 6.2. Resin production facility.

follows. Five components are charged to one of the reactors R01 to R04. When reaction starts, the dump time is predicted using a model and inputting data such as component quantities and reaction temperature. Based on expected dump time and quantities charged, the processing rate to a stripping tower is gradually adjusted in such a way that the dump unit has always enough room to receive the reactor dumps but never has such a low level that the tower has to be shut down. When the dumping unit receives a batch or has a high level, its "availability" flag is set to zero; when it is not active or can accept another reactor dump, the flag is set to one. Reactors are served on a first-come first-served basis. The dump unit maintains a table with expected dump times for each reactor. It should be mentioned that all four reactors are identical, which justifies the first-come first-served queuing mechanism.

FUNCTIONS OF A BATCH CONTROL PACKAGE

The functions of a batch control package can be summarized as follows:[29,30]

- Discrete (on/off) control. This deals with the checking of contact inputs and the driving of contact outputs. When there is a discrepancy between the actual state of a contact and its desired state, one or more actions can be taken:

  1. generating an alarm message
  2. invoking an alarm/service subroutine which has been coded by the user and deals with the discrepancy
  3. shutting down part or all of the process

- Interface with the continuous control package. For example, a blower is started (contact-output activated) upon low flow (analog value).

- Sequence control (start/stop pumps, open/close valves, etc.). Sequence control is usually handled by the batch executive, the main program that controls the execution of other programs and interfaces with status tables which contain process unit information related to the batch type of operation.

- Performance of calculations

- Notifying operator of alarm conditions

- Providing an operator interface to the process

- Reporting and logging

- Recipe handling

When a large variety of products are manufactured in the same process facility, recipe handling is required. The recipe is a set of operating variables for a particular batch or grade of product and can include feed quantities, temperatures, pressures, levels, and also control loop settings, for example, if they are product dependent. Examples are given in Table 6.1.

Often in batch process control, "jargon" or a special terminology is used. Some of the most common expressions are given below:

- A "unit" is part of the process related to a specific operation, e.g., a distillation tower or a reactor.

- "Shared equipment" is equipment that is used by more than one unit.

- A "phase" is part of a production sequence, e.g., a heat-up phase.

A unit can be running in different operating modes. Different batch control systems use a different terminology, although the following modes are common:

- "automatic." The sequence is carried out under control of a user-written program without operator intervention. This is the normal operating mode.

- "manual." Process action is carried out by the operator from and via the operator interface.

Table 6.1.   Recipes for Different Product Grades

Recipe	Min	Max	Actual
	001 Grade A		
Amount of component, KG	4000.0	6000.0	4745.0
Reaction temperature, °C	100.0	120.0	114.7
Pressure, KPA	20.0	90.0	55.0
Level, PCT	80.0	90.0	82.0
Temp. controller gain	10.0	20.0	12.5
Deriv action, min	0.0	2.0	0.08
	002 Grade B		
Amount of component KG	4000.0	6000.0	5250.0
Reaction temperature, °C	100.0	120.0	118.5
Pressure, KPA	20.0	90.0	85.5
Level, PCT	80.0	90.0	87.0
Temp. controller gain	10.0	20.0	16.0
Deriv action, min	0.0	2.0	0.06

Depending on the batch control package, more operating modes may exist. When a unit runs in automatic mode, it can be in different states:

- "normal." In this state, sequence control runs without a problem.

- "hold." In this mode, sequence control is halted as a result of predefined error conditions, and the unit is brought to a predefined safe state.

Different batch control packages may use one or more additional states. The way table-driven batch control software works is as follows: the executive scans the status tables to see if any unit needs to be serviced. The status tables contain all information regarding the state of a unit. The unit is not serviced if it is waiting or in manual. A phase which requires an output to be modified will set up a request via a utility routine in the input/output tables. The actual change of the output is carried out by a program which manages all input/output communication. The consistency between the required output state and the actual state is continuously verified by the input/output consistency check program, and, if a discrepancy occurs, the program will initiate the hold request and the appropriate alarms.

All alarms and messages are usually set up as requests in a message queue, and an independent alarm and message handler ensures that this information is presented to the operator via the appropriate interface.

BENEFITS OF BATCH CONTROL

The major benefit of automation of batch processes is increased safety. Because of high-frequency (typically two seconds or less) scanning of contact inputs, such as the status of all valves and pumps in the process, a discrepancy will be determined immediately. Upon detection of a discrepancy, predefined action will be taken, and, in case of an emergency, the process will be brought to a safe state. Possible errors in manual actions are eliminated, leading to increased personnel and plant safety.

It is difficult to convert increased safety into an economic gain in terms of k$/year. It is sometimes necessary to operate under computer control for two or more years before a real comparison with past operation without a computer can be made and exact credits can be identified.

The second most important benefit of automation of a batch process is increased production through a reduction in cycle time. Considerable savings are possible, especially when many steps are involved. The following savings were calculated for a batch plant with four reactors, for which the following data were available:

- cycle time reactors reduced from 12 hours to 10.5 hours

- average number of operating days per year 330

- incremental value (profit) of a reactor batch $2,400 which results in savings of:

$$(24/10.5 - 24/12) * 4 * 330 * 2.4 = 905 \text{ k\$/year}$$

Cycle times can usually be reduced by 10% to 15% through automation.[27]

Another important benefit is increased consistency of product quality. In a competitive environment, it is extremely important to produce a consistent product. In a decade of overproduction, quality is more important than quantity. Off-spec incidents can usually be reduced by a factor of four, which can be translated into an economic gain.

A last benefit that should be mentioned is flexibility in charging through the use of multiple recipes. Especially when many different products are made, a situation exists which is prone to errors. By having a recipe for each product grade, the required ingredients are added in appropriate amounts, react at the right temperature, and so forth.

IMPLEMENTATION AND MAINTENANCE OF BATCH CONTROL SOFTWARE

The first requirement for successful implementation of batch control is that the process and the various operations are well-defined. Batch control logic is more complicated than a continuous control application, and, therefore, changes to the program logic will be more time consuming than changes to a control strategy in a continuous process environment. Testing of any control application will require that the necessary safety precautions are taken since changes to the process may be major, e.g., pumps may start or valves may open. In a continuous control application, setpoints of secondary loops are usually changed gradually; hence, after a few program executions, an idea can be obtained about how well the control program is working. Therefore, continuous control applications can be tested on-line as a rule, whereas in batch control, certain parts of the process may have to be isolated for test purposes. Testing of a batch control application will require a team environment where process operator, plant engineer, and control engineer work together, and all have a thorough understanding of the steps involved in the testing and understand possible health and safety hazards in case the test fails. The time required for the testing of a batch application should not be underestimated. Often four or five test runs are required before the entire application runs successfully. This number is not absolute; much depends on the experience of the control team, the state and the reliability of the hardware, and so forth.

Since much depends on contact inputs and outputs, switches should be extremely reliable. Nothing is more annoying than switches that do not function properly and trip to invoke other programs (e.g., drive to a safe state).

The maintenance of batch control programs requires more effort than the maintenance of a continuous control application. Usually any continuous control application runs maintenance free once it is thoroughly tested and commissioned. A batch environment is more prone to process, equipment, and instrument changes, which will all require program changes. After these changes are made and implemented, another test run may be required to check the modified control sequence. An important aspect of efficient maintenance is program documentation. Since the design engineer often leaves the project after two to three years, other people have to make changes to programs they did not design. A good practice is to use a program heading which gives the dates of the original design and design revisions, the name of the engineer who made the design c.q. revision, a short description of each change, and a functional description of the application. All variables that are used in the control program should be defined. The control program can best be divided into blocks which each have a specific function; further detailed documentation can be done at the beginning of each block. A well-defined control program is not only easier to maintain, but it also increases safety because it reduces the risk of making errors.

CONTINUOUS CONTROL

In continuous control, process variables are usually maintained at their setpoints. Disturbances will cause the process variables to drift away from setpoint, making control necessary in order to bring the process variables back to their setpoints. The control of single input-single output systems is done using microprocessor-based controllers (basic control). The next level of control, however, can create a more complex control scheme using multiple inputs and multiple outputs, the latter usually being the setpoints of the basic controllers.

Simple, single input-single output controllers often use a PID control algorithm:

$$u_k = u_{k-1} + K_c [e_k - e_{k-1} + \frac{\Delta t}{\tau_i} e_k + \frac{\tau_d}{\Delta t} (e_k - 2e_{k-1} + e_{k-2})] \qquad (6.1)$$

in which u_k = control signal at time k
K_c = controller gain
τ_i = integral time, min
τ_d = derivative time, min
Δt = controller execution interval, min
e_k = setpoint – process value at time k

Most PID controllers in modern instrumentation systems use equation (6.1) in calculating the controller output. A feature that is added is the so-called reset wind-up prevention. When the valve reaches one of its limits (0 or 100 percent), the integral action is eliminated in order to prevent the controller to calculate a new output continually. Many techniques can be applied to include anti-reset wind-up in a control strategy[31] (see chapter 12).

MULTIVARIABLE SYSTEMS[32]

The PID controller uses the process input as the control variable and the process output as the controlled variable. In a process with multiple inputs and multiple outputs, however, one will usually not find a situation where one control variable can be paired with one controlled variable without interactions from other control variables.

A simple example is the blending of one component with an inert flow. If the concentration in the final blend changes, the flow of the component can be changed in order to control the composition. However, this will change the total flow of blended material, which requires the inert flow to be changed as well. The control problem cannot simply be treated as two independent control loops. A number of techniques is available to solve the problem (Figure 6.3a; Chapter 11). By applying the technique shown, independent control loops are obtained. In the literature, this type of control is often called "noninteractive control." The major advantage of this approach is that the control scheme is visible to the operator since it is broken down into independent control loops. Failure of one control loop does not affect the rest of the control strategy.

Another approach is to use "interactive control" (Figure 6.3b; Roffel et al.[33]). All process outputs now have an impact on all process inputs. This type of control is more complicated from an operator point of view. The method requires a process model, giving the relationship between all process inputs and outputs (state variables):

$$u_{1,k} = a_{11}x_{1,k-1} + a_{12}x_{2,k-1} + \ldots + b_{11}u_{1,k-1} + b_{12}u_{2,k-1} + \ldots$$

$$u_{2,k} = a_{21}x_{1,k-1} + a_{22}x_{2,k-1} + \ldots + b_{21}u_{1,k-1} + b_{22}u_{2,k-1} + \ldots$$ \hfill (6.2)

in which $x_1 \ldots x_n$ state variables
$u_1 \ldots u_n$ control variables

Equation (6.2) can be written in matrix notation as:

$$x_k = A\,x_{k-1} + B\,u_{k-1}$$ \hfill (6.3)

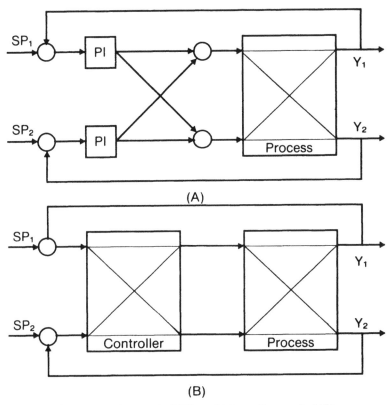

Figure 6.3. Noninteractive control (A) and interactive control (B).

The process model is usually combined with a measurement model:

$$y_k = C\, x_k \qquad (6.4)$$

in which y is the measurement vector. The feedback control law can be written as:

$$u_{k+1} = G\, y_k \qquad (6.5)$$

in which A, B, C, and G are matrices.

Optimal control theory can be applied using a quadratic cost function. The result is a calculation of the gain coefficients in the gain matrix G. The process description can easily be modified to include process and measurement noise. The solution for the gain matrix G depends strongly on these estimates as well as on the weighing coefficients in the quadratic cost function.

OTHER CONTROL TECHNIQUES

Several other techniques to improve continuous control are suitable for implementation at the supervisory control level. Some of these include dead time compensation, feed-forward control, and adaptive control. These techniques will be discussed in detail in the following chapters. With respect to implementation and documentation of continuous control programs, the same holds for batch control systems, as has been stated.

CHAPTER 7

Introduction to Advanced Control

Ninety percent of the control loops in the chemical and petrochemical industry work well by using simple proportional, integral, and derivative (PID) control. The PID controller has been used for decades, first in pneumatic instrumentation, later in analog instrumentation, and today in digital instrumentation and computer control systems.

In some cases the PID controller does not perform so well, and it may be necessary to study and analyze these situations. Personal computers offer the control engineer an excellent tool to study the behavior of controlled and uncontrolled processes. Several computer-aided design packages with tremendous flexibility to analyze virtually anything the control engineer can think of are on the market.

This chapter, however, will familiarize the reader with using Lotus 1-2-3* as a simple but efficient tool to simulate and analyze simple control problems. The major advantage of this approach is that once one is able to define and solve the problem this way, one can easily develop a control application on any process computer. For this and the following chapter, examples have been worked out on the diskette using Lotus 1-2-3.

COMMON ELEMENTS IN PROCESS DYNAMICS

When a process is disturbed at the inlet, the outlet will not change immediately but will usually slowly move toward a new steady state value. This can be illustrated by a simple example. Consider a tank with an inflow of water at rate F and temperature T_i, as shown in Figure 7.1. If there is no generation or loss of energy, balance can be written as:

$$\rho V c_p \frac{dT_0}{dt} = F c_p T_i - F c_p T_0 \tag{7.1}$$

---

*Lotus and 1-2-3 are registered trademarks of Lotus Development Corporation.

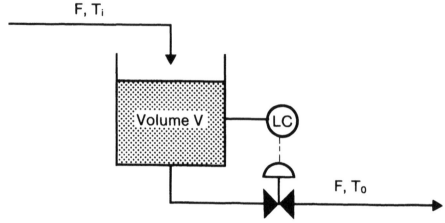

Figure 7.1. Tank with level control.

in which ρ = liquid density, kg/m³
V = tank volume, m³
c_p = specific heat, J/kg°C
T = temperature, °C
F = mass flow, kg/min

Equation (7.1) can be rewritten as:

$$\tau \frac{dT_0}{dt} = T_i - T_0 \tag{7.2}$$

in which $\tau = \rho V/F$, the residence time in the tank. The solution of this equation is given in many textbooks (Roffel and Rijnsdorp[26]) as:

$$T_0 = T_{0,ss} + (T_i - T_{i,ss})(1 - e^{-t/\tau}) \tag{7.3}$$

in which the subscript *ss* indicates steady state. The response to a step change in T_i from 30°C to 40°C is shown in Figure 7.2.

Control engineers often write an expression like equation (7.2) as a transfer function. A transfer function determines the relationship between process input and output. Define deviation variables as:

$$\delta T_0 = T_0 - T_{ss}$$

$$\delta T_i = T_i - T_{ss} \tag{7.4}$$

T_0, °C   40

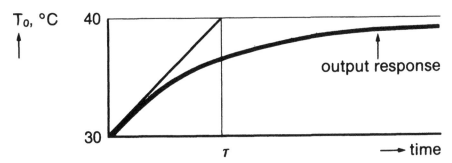

output response

30

τ

time

Figure 7.2.   Response to a step change.

T_i

T_0

Figure 7.3.   Flow through a pipe.

then equation (7.2) can be written as:

$$\tau \frac{d(\delta T_0)}{dt} = \delta T_i - \delta T_0 \qquad (7.5)$$

Introducing the operator s = d/dt, equation (7.5) can be written as:

$$\frac{\delta T_0}{\delta T_i} = \frac{1}{\tau s + 1} \qquad (7.6)$$

This expression is called a *first order* transfer function.

Another transfer function encountered in the process industry is a so-called dead time. Figure 7.3 shows a dead time process. Liquid enters a long pipe at temperature T_i and leaves the pipe at temperature T_0. If no heat loss or heat generation takes place, the transfer function becomes:

$$\frac{\delta T_0}{\delta T_i} = e^{-\theta s} \qquad (7.7)$$

in which θ is the time for the liquid to travel from inlet to outlet. The step response is shown in Figure 7.4.

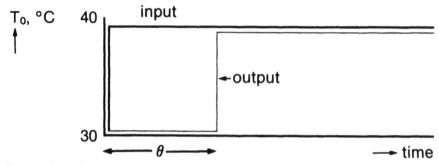

Figure 7.4.   Step response of dead time process.

COMMON ELEMENTS IN CONTROL

As mentioned before, the PID controller is commonly used in industry. The relationship between input and output of the controller can be given by:

$$\frac{\delta u}{\delta e} = K_c[1 + \frac{1}{\tau_i \cdot s} + \tau_d \cdot s] \tag{7.8}$$

in which δu = change in controller output
$\quad\quad\quad \delta e$ = error, setpoint – process value
$\quad\quad\quad K_c$ = controller gain
$\quad\quad\quad \tau_i$ = integral action
$\quad\quad\quad \tau_d$ = derivative action
$\quad\quad\quad s$ = d/dt operator

Another element, often used in feed-forward control, is the so-called lead-lag element. The transfer function of this element is:

$$\frac{\delta y}{\delta x} = K \frac{1 + \tau_1 s}{1 + \tau_2 s} \tag{7.9}$$

with δy = output change
$\quad\quad \delta x$ = input change
$\quad\quad \tau_1$ = lead time constant
$\quad\quad \tau_2$ = lag time constant
$\quad\quad K$ = gain

The step response of this element for $K = 1$ is shown in Figure 7.5. If $\tau_1/\tau_2 > 1$, the output peaks initially as shown. If $\tau_1/\tau_2 < 1$, the output jumps to the value τ_1/τ_2 at time zero and then will approach the input via a first order response.

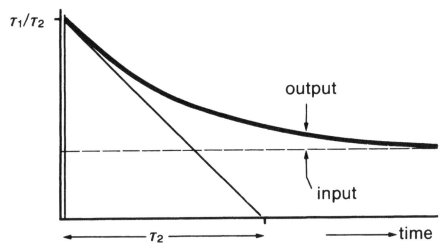

Figure 7.5. Lead-lag response.

DISCRETE APPROXIMATIONS

The transfer functions shown in the previous sections used the Laplace operator s, which offered a suitable way of representing continuous systems. However, today's computers are digital computers which can make recursive calculations, and, therefore, the transfer functions will be discretized. In order to do this, s is replaced again by d/dt. In the first order equation the derivative is replaced by:

$$\frac{dT_0}{dt} = \frac{T_{0,k} - T_{0,k-1}}{\Delta t} \tag{7.10}$$

Equation (7.2) can then be rewritten as:

$$\frac{\tau}{\Delta t}\,[T_{0,k} - T_{0,k-1}] = T_{0,k} - T_{i,k} \tag{7.11}$$

which can be written as:

$$T_{0,k} = KLAG* \, T_{0,k-1} + (1 - KLAG)* \, T_{i,k} \tag{7.12}$$

with

$$KLAG = \tau/(\tau + \Delta t) \tag{7.13}$$

$$\Delta t = \text{discretization interval}$$

The approximation of equation (7.10) is accurate as long as Δt is small compared to τ. For larger values of Δt a better approximation is:

$$\text{KLAG} = e^{-\Delta t/\tau} \tag{7.14}$$

In order to determine how accurate the discrete approximation is, in Table 7.1 the solution of the discrete equation is compared to the exact solution as given in equation (7.3). This is done for a value of $\tau = 5.0$.

It can be seen that the discrete equation introduces some error. However, as a rule of thumb, Δt is usually chosen equal to:

$$\Delta t = (\tfrac{1}{3} \ldots \tfrac{1}{5})\tau \tag{7.15}$$

in which case the discretization error is rarely more than 5%. If the model would be used in control, the control scheme must be able to cope with model errors of at least 25%. Therefore one does not have to be concerned about some model error and can use equations (7.12) and (7.13) as the discrete version of equation (7.2).

The next dynamic element that was of interest was the dead time element. The discrete version of the element (equation 7.7) is:

$$T_{0,k} = T_{i,k-\theta/\Delta t} \tag{7.16}$$

The discrete version of a PID controller can easily be derived from equation (7.8) which can be written as:

Table 7.1. Discrete Approximation of First Order Equation

Time (min)	T_i	Exact solution for T_0 acc. to eqn. (7.3)	T_0 using eqn. (7.13) & (7.14) $\Delta t = 1$	T_0 using eqn. (7.13) & (7.14) $\Delta t = 2$
0	30	30.00	30.00	30.00
1	40	31.81	31.67	
2	40	33.30	33.06	32.86
3	40	34.51	34.22	
4	40	35.51	35.19	34.90
5	40	36.32	35.99	
6	40	36.99	36.66	36.36
7	40	37.53	37.22	
8	40	37.98	37.68	37.40
9	40	38.35	38.07	
10	40	38.65	38.35	38.14

$$\tau_i \cdot s \cdot u = (K_c \cdot \tau_i \cdot s + K_c + K_c \cdot \tau_i \cdot \tau_d \cdot s^2)e \qquad (7.17)$$

Now substitute:

$$s \cdot u = \frac{du}{dt} = \frac{u_{k+1} - u_k}{\Delta t} \qquad (7.18)$$

$$s \cdot e = \frac{de}{dt} = \frac{e_k - e_{k-1}}{\Delta t} \qquad (7.19)$$

$$s^2 \cdot e = \frac{d^2u}{dt^2} = \frac{e_k - 2e_{k-1} + e_{k-2}}{\Delta t^2} \qquad (7.20)$$

which gives as the discrete version of the PID controller:

$$u_{k+1} = u_k + K_c[e_k - e_{k-1} + \frac{\Delta t}{\tau_i} \cdot e_k + \frac{\tau_d}{\Delta t} \cdot (e_k - 2e_{k-1} + e_{k-2})] \qquad (7.21)$$

Similarly the discrete equation for a lead-lag element can be derived:

$$y_k = \underbrace{KLAG \cdot y_{k-1} + K(1 - KLAG) x_k}_{\text{Lag Part}} + \underbrace{KLEAD(x_k - x_{k-1})}_{\text{Lead Part}} \qquad (7.22)$$

in which $\quad KLAG = \tau_2/(\tau_2 + \Delta t)$

$$\qquad (7.23)$$

$$KLEAD = K \, \tau_1/(\tau_2 + \Delta t)$$

In case the first order transfer function has a gain K, the discrete approximation of the first order transfer function becomes:

$$y_k = KLAG \cdot y_{k-1} + K(1 - KLAG)x_k \qquad (7.24)$$

Equations (7.16), (7.24), (7.22) and (7.23) will be frequently used in control strategies.

z-TRANSFORM

Many modern textbooks[34] use the z-transform as a convenient way of representing the transfer function of discrete systems. The z-transform can easily be derived from the discrete equation by introducing the backward shift operator z^{-1}:

$$x_{k-1} = z^{-1} x_k \tag{7.25}$$

Introduction of (7.24) into (7.25) gives:

$$y_k = KLAG \cdot z^{-1} \cdot y_k + K(1 - KLAG)x_k \tag{7.26}$$

The sample/hold system always introduces one additional period of delay which has to be accounted for in the process transfer function. Equation (7.26) can therefore be written as:

$$\frac{y_k}{x_k} = \frac{K_p z^{-1}}{1 - dz^{-1}} \tag{7.27}$$

with

$$d = KLAG$$
$$K_p = K(1 - d) \tag{7.28}$$

The z-transformation of the dead time function is:

$$\frac{y_k}{x_k} = z^{-\theta/\Delta t - 1} \tag{7.29}$$

which can easily be seen from equation (7.16).
   Equation (7.21) gives for a PID controller:

$$\frac{y_k}{x_k} = \frac{a + bz^{-1} + cz^{-2}}{1 - z^{-1}} \tag{7.30}$$

with

$$a = K_c[1 + \frac{\Delta t}{\tau_i} + \frac{\tau_d}{\Delta t}]$$

$$b = -K_c[1 + \frac{2 \cdot \tau_d}{\Delta t}] \tag{7.31}$$

$$c = \frac{\tau_d}{\Delta t}$$

Substitution of equation (7.25) into (7.22) gives for the transfer function of a lead-lag in z-notation:

$$\frac{y_k}{x_k} = \frac{K^*(1 - ez^{-1})}{1 - dz^{-1}} \tag{7.32}$$

with

$$e = \frac{KLEAD}{K(1 - d) + KLEAD}$$

and   (7.33)

$$K^* = K(1 - d) + KLEAD$$

CONTROL SYSTEM SIMULATION

In this section the simulation of a simple control system using Lotus 1-2-3 will be illustrated. Consider the system shown in Figure 7.6 which is frequently encountered in the process industry. The process in z-notation is shown in Figure 7.7. The values of a . . . d and K_p are given in the previous section.

In order to get some insight into control of this system, one can simulate it using Lotus 1-2-3 for $\tau = 4$ min, $\theta = 1$ min, $K = 0.025$ and $\Delta t = 1$ min. From equation (7.23) KLAG can be calculated; in this case it is equal to 0.8. If the steady state process input u equals 170.0 and the steady state process output y equals 4.25, the process equation in terms of deviation variables becomes:

$$y_k = 0.8 \ (y_{k-1} - 4.25) + 0.025* \ 0.2* \ (u_{k-2} - 170) + 4.25$$

The setup for the simulation of the controlled process is shown in Table 7.2. Data entries can be made in the following fields: B6, C6, and D6 for K_c, τ_i, and τ_d respectively and G7 for a change in setpoint. The /COPY command can be used to copy the cell contents to other rows. The F10 key can be used to call up the display once it has been fully defined. In case of a system with ten minutes dead time, cell F6 is copied to cell F16 and the contents of cell F17 become:

$$F17 = 0.8* \ (F16 - 4.25) + 0.025* \ 0.20* \ (E6 - 170) + 4.25$$

The diskette contains two files, PID and PIDDT; the first simulates a process with a one-minute dead time, the second simulates a system with a ten-minute dead time. By changing K_c, τ_i, and τ_d, PID control of a first order/delay system can be studied.

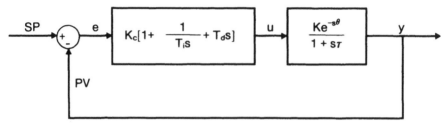

Figure 7.6.   PID control of 1st order/delay system in s-representation.

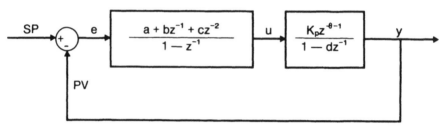

Figure 7.7.   PID control of 1st order/delay system in z-representation.

Table 7.2. **Setup of Spreadsheet for Control System Simulation**

	Time A	K_c B	τ_i C	τ_d D	Process Input u_k E	Process Output y_k F	sp_k G	Error H	u_k I
6	0.	25.	1.	0.	170.	4.25	4.25	[h]	0.
7	[a]	[b]	[c]	[d]	[e]	4.25	4.25		0.
8						[f]	[g]		[i]
9									
.									
.									

Note: Contents of the cells are:

[a]A7 = +A6 + 1
[b]B7 = +B6
[c]C7 = +C6
[d]D7 = +D6
[e]E7 = +E6 +I6
[f]F8 = 0.8 • (F7–4.25) + 0.025 • 0.2 • (E6–170) + 4.25   process equation
[g]G8 = +G7
[h]H6 = +G6–F6
[i]I8   = +B8 • [(H8–H7 + (1/C8) • H8 + D8 • (H8–2 • H7 + H6)]   controller equation

Table 7.3.   PID Tuning Guidelines

Loop Type	Controller Gain	Integral Time (τ_i) (min)	Derivative Time (τ_d) (min)
Flows	0.3	0.0625	0.0
Levels, fast	1.0	8.0	0.0
Levels, slow	0.25	16.0	0.0
Pressures, fast	2.0	0.5	0.0
Pressures, slow	1.0	2.0	0.125
Temperatures, fast	1.0	2.0	0.0
Temperatures, slow	1.0	16.0	0.250
Compressor speed	1.0	2.0	0.0

PID TUNING GUIDELINES

During the past decades many guidelines for tuning controllers have been developed. Table 7.3 can be used as a rule of thumb. With the guidelines given in Table 7.3 and the two simulations PID and PIDDT, which you can modify to suit your needs, you should be able to tune most control loops.

The next chapter will deal with dead time compensators as a means to improve control performance of systems with large time delays. Three techniques for dead time compensation will be discussed.

CHAPTER 8

Dead Time Compensators

One of the most common control problems in the process industry today is control of a process with significant dead time. Control of such systems will usually involve a model-based approach. The two techniques that will be discussed in this chapter are the Smith Predictor and the Dynamic Reconciliator dead time compensators. These will be applied to SISO (Single Input Single Output) systems and in subsequent chapters the problem of feed-forward, decoupling, and a fully interactive 2×2 system will be addressed. The self-tuning regulator (STR), which has rapidly gained in popularity in recent years and which not only deals with the problem of dead time but also changing process parameters, e.g., process gain of a fixed-bed reactor changes with the deactivation of the catalyst, will also be discussed briefly in this chapter and in more detail in Chapter 15.

Dead time compensator algorithms are typically written at the supervisory (computer) level and output to the basic regulatory level. In other words, they are usually applied to cascade systems. Generally the process model is assumed to be first order with dead time although this is not a restriction to the applicability of the dead time compensation techniques. However, this type of process model is simple and is suitable for many processes in the chemical and petrochemical industry. As a rule of thumb, dead time compensation techniques are applied whenever the dead time exceeds two times the time constant ($\theta \geq 2\tau$).

SMITH PREDICTOR [35]

The control block diagram for a SISO Smith Predictor control strategy is shown in Figure 8.1. More specifically, this is an example of using the reflux flowrate to control the impurities in the overhead of a distillation tower via an analyzer feedback loop. Dead time can arise e.g. from analyzer cycling time, holdup in the reflux drum (analyzer located on the reflux or distillate line), and so forth.

In Figure 8.1, the Smith Predictor *predicts* the concentration, ($c_{k+\theta}$), θ time slots ahead and subtracts the present predicted concentration c_k. This value is then subtracted from the *actual* analyzer reading c resulting in the compen-

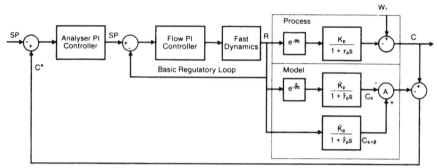

Figure 8.1. Smith Predictor control block diagram. w_1 = load disturbance; K_p = process gain; τ_p = process time constant; θ = process time delay; $\hat{K}_p$ = estimated process gain; $\hat{\tau}_p$ = estimated process time constant; $\hat{\theta}_p$ = estimated process time delay; R = reflux flow; and c = concentration of the impurities.

sated value c* that will be used in the analyzer PI controller. The predicted concentration $c_{k+\theta}$ is calculated using:

Laplace Transform:

$$c_{k+\theta} = \frac{\hat{K}_p}{1 + \hat{\tau}_p s} \cdot R_k \qquad (8.1a)$$

When we discretize equation (8.1a) and use absolute values for $c_{k+\theta}$ and R_k the result will be:

Discretized Form:

$$c_{k+\theta} = \text{KLAG} \cdot c_{k+\theta-1} + (1 - \text{KLAG}) \cdot \hat{K}_p \cdot R_k + \text{bias} \qquad (8.1b)$$

or

z-Transform:

$$c_{k+\theta} = \frac{(1-\text{KLAG})\hat{K}_p \cdot R_k z^{-1}}{1 - \text{KLAG} \, z^{-1}} + \frac{\text{bias}}{1-\text{KLAG} \, z^{-1}} \qquad (8.1c)$$

where
$c_{k+\theta}$ = impurity concentration predicted θ time slots ahead
$\text{KLAG} = \hat{\tau}_p/(\hat{\tau}_p + \Delta t)$
Δt = sampling time
R_k = present reflux flowrate
z^{-1} = backward shift operator, $c_{k-1} = z^{-1}c_k$
$\text{bias} = (c_{ss} - K \, R_{ss}) * (1 - \text{KLAG})$

c_{ss} = impurities concentration at steady state
R_{ss} = reflux flowrate at steady state

The present predicted concentration c_k is calculated according to:

Laplace Transform:

$$c_k = \frac{\hat{K}_p \cdot e^{-\hat{\theta}s}}{1 + \hat{\tau}_p s} \cdot R_k \qquad (8.2a)$$

or

Discretized Form:

$$c_k = KLAG \cdot c_{k-1} + (1 - KLAG) \cdot K_p \cdot R_{k-\theta} + bias \qquad (8.2b)$$

or

z-Transform:

$$c_k = \frac{(1 - KLAG)\hat{K}_p \, z^{-\hat{\theta}/\Delta t - 1}}{1 - KLAG \, z^{-1}} \cdot R_k + \frac{bias}{1 - KLAG \, z^{-1}} \qquad (8.2c)$$

The dynamically compensated concentration c* can be calculated from the foregoing equations by subtracting equation (8.2) from (8.1) and then subtracting the actual impurity concentration $c_{k,actual}$:

Laplace Transform:

$$c^* = c_{k,actual} - \left[\frac{\hat{K}_p}{1 + \hat{\tau}_p s} \cdot R_k(1 - e^{-\hat{\theta}s}) \right] \qquad (8.3a)$$

or

Discretized Form:

$$c^* = c_{k,actual} - [KLAG \cdot (c_{k+\theta-1} - c_{k-1}) + (1 - KLAG) \cdot \hat{K}_p \cdot (R_k - R_{k-\theta})] \qquad (8.3b)$$

or

z-Transform:

$$c^* = c_{k,actual} - \left[\frac{(1-KLAG)\hat{K}_p z^{-1}}{1 - KLAG \, z^{-1}} [1 - z^{-\hat{\theta}/\Delta t}]R_k \right] \qquad (8.3c)$$

Because the concentration of the impurities is dynamically compensated, the analyzer PI controller can be tightly tuned without producing the oscillatory response typical of noncompensated feedback loops with large dead time.

Whenever the Laplace transform of a transfer function is discussed, input and output measurements are assumed to be expressed in terms of deviation variables (absolute value minus steady state value). If absolute values are used, which is highly desirable from a practical viewpoint, then an additional term (bias) will be introduced in the discretized equations (8.1b), (8.1c), (8.2b), and (8.2c). However, it is not necessary to determine its value since it is cancelled at the summation point A, shown in Figure 8.1.

There are many ways in which the Smith Predictor block diagram can be equivalently represented. Another simpler representation is shown in Figure 8.2. However, implementing the control strategy as shown in Figure 8.1 allows the user to "fine-tune" the process model parameters on-line by comparing c_k against the actual analyzer reading. Once this comparison is favorable, the control engineer is assured of a good process model and thus good feedback control. The analyzer PI controller can then be tuned as if no dead time exists in the analyzer feedback loop.

Although a first order transfer function plus dead time model is assumed, the Smith Predictor technique is by no means limited to this special case. The method holds equally well for higher order models plus dead time. A first order plus dead time model is shown because most processes in the process industry can be adequately described by such a model. It should be noted that the signs (+ and –) shown in the summation junctions of Figures 8.1 and 8.2 are specific only to the preceding example. Different situations will require different summation junctions, requiring each situation to be analyzed in detail. As stated before, the analyzer PI controller can be tuned as if the process had no dead time. The following tuning guideline works well: controller gain K_c equal to the inverse of the process gain K_p and the integral time equal to the process time constant τ_p.

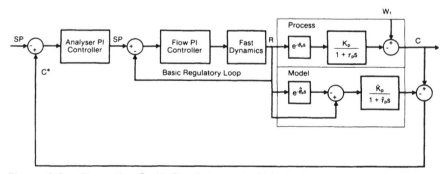

Figure 8.2. Alternative Smith Predictor control block diagram.

DYNAMIC RECONCILIATOR

This technique, originally proposed by Bartman,[36] also uses a first order transfer function plus dead time model:

$$\frac{Y}{U} = \frac{Ke^{-\theta s}}{1 + \tau s} \qquad (8.4)$$

or

$$Y = KGU$$

where $\quad G = \dfrac{e^{-\theta s}}{1 + \tau s}$

θ = process dead-time
τ = process time constant
K = process gain
Y = output deviation variable = $y - y_{ss}$
y = absolute value of output
y_{ss} = steady-state value of the output
U = input deviation variable = $u - u_{ss}$
u = absolute value of input
u_{ss} = steady-state value of the input

As noted previously, for absolute values, equation (8.4) then becomes:

$$y = KGu + bias \qquad (8.5)$$

where bias = $(y_{ss} - Ku_{ss})(1 - KLAG)$ if equation (8.5) is discretized via a backward difference approximation. KLAG is as defined before. The bias is not calculated directly but is estimated from (see equation [8.5]):

$$bias = y - KGu \qquad (8.6)$$

The control equation is derived from the static version of equation (8.5):

$$u_{sp} = \frac{y_{sp} - bias}{K} \qquad (8.7)$$

which leads to the SISO control structure shown in Figure 8.3 in which

u = absolute value of the manipulated variable (e.g., flow or temperature)
y = absolute value of the controlled variable (e.g., concentration or conversion)

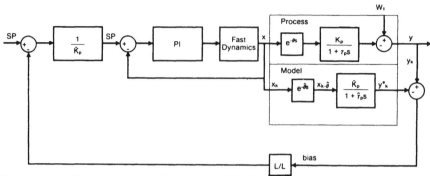

Figure 8.3. Dynamic reconciliator control block diagram.

L/L = Lead-lag block with gain = 1

With reference to Figure 8.3 and equation (8.6):

$$\text{bias} = y_k - y^*_k$$

where   y = *absolute* value of the output
         y* = model output using the *absolute* value as the input

The value of y* is calculated as follows:

$$y^* = \frac{K_p e^{-\theta s}}{1 + \tau_p s} \cdot u$$

which can be written as:

$$y^* + \tau_p \left(\frac{dy^*}{dt} \right) = K_p \cdot u(t-\theta)$$

or, when using the backward difference approximation:

$$y^*_k = y^*_{k-1} \cdot KLAG + (1 - KLAG) \cdot K_p \cdot u_{k-\theta/\Delta t-1} \tag{8.8}$$

where $KLAG = \tau_p/(\tau_p + \Delta t)$
       Δt = sampling time
       $u_{k-\theta/\Delta t-1}$ = absolute value of input that has been delayed by θ units of time

The lead-lag block in the feedback path allows the user to fine-tune the feedback controller. For example, if the process output (y_k) is noisy, a first order lag would be appropriate. On the other hand, if disturbances have slow dynamics, a lead-lag with more lead than lag would be appropriate. Extensive simulations gave the following guidelines for tuning the lead-lag:

$$A = \text{MAX}(\theta, \tau) \tag{8.9}$$

$$B = \text{MIN}(\theta, \tau) \tag{8.10}$$

$$\tau_{\text{lag}} = 0.5 \, A \tag{8.11}$$

$$\tau_{\text{lead}} = 0.5[A + (2-\theta/\tau)B] \tag{8.12}$$

It was found that, in many cases, control performance is satisfactory without using the feedback lead-lag.

As in the case with the Smith Predictor, process model parameters can be fine-tuned on-line by comparing y^*_k with actual output y_k. Also, the signs (+ and –) shown in the summation junctions of Figure 8.3 are specific to a process where an increase in the input x will result in an increase in the output y (positive gain).

Processes with higher order model plus dead time are handled in a similar manner. The only difference is in the form of the discretized equation (8.8) that is used.

SELF-TUNING REGULATORS

Although self-tuning regulators will be dealt with in detail in a later chapter, a brief summary will be given here since they can also be used in model-based control in which the dead time has to be eliminated.

Self-tuning regulators (STR) can be particularly useful in processes with changing parameters, e.g., fouling of equipment or decay of catalyst. The process may have a dead time, however the value of it should not change. A self-tuning regulator has the structure as shown in Figure 8.4. In this technique a process model is assumed, and a recursive least squares estimation algorithm is used to update the model parameters on-line based on current and past

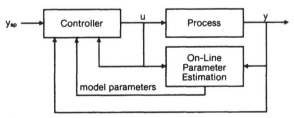

Figure 8.4. Control using self-tuning regulator.

process input and output values. After the model parameters are updated, the new control output is calculated.

Example:

Assume that a particular process model can be described by

$$\hat{y}_k = \alpha_0 y_{k-2} + \beta_0 \nabla u_{k-2} + \beta_1 \nabla u_{k-3} \qquad (8.13)$$

in which $\quad \hat{y}_k$ = predicted deviation betwen process output and setpoint
y_{k-2} = measured deviation between process output and setpoint, two time steps ago
$\nabla u_{k-2} = u_{k-2} - u_{k-3}$
u_{k-2} = process input two time steps ago
u_{k-3} = process input three time steps ago
α, β = model parameters

This model is now used to find the prediction $\hat{y}_{k+2}$:

$$\hat{y}_{k+2} = \alpha_0 y_k + \beta_0 \nabla u_k + \beta_1 \nabla u_{k-1} \qquad (8.14)$$

The target is to control or setpoint, hence the best estimate is $\hat{y}_{k+2} = 0$, therefore:

$$\alpha_0 y_k + \beta_0 \nabla u_k + \beta_1 \nabla u_{k-1} = 0 \qquad (8.15)$$

resulting in:

$$u_k = \left(1 - \frac{\beta_1}{\beta_0}\right) u_{k-1} + \frac{\beta_1}{\beta_0} u_{k-2} - \frac{\alpha_0}{\beta_0} y_k \qquad (8.16)$$

which shows that the current control input depends on the current process output and the previous two control inputs. The parameters α_0, β_0, and β_1 can be updated via a least squares estimation technique.

OTHER ALGORITHMS

Badavas[37] proposed a so-called direct synthesis controller (DSC) for control of processes with dead time. The process output is fed directly into the controller; the controller output adjusts the process input.

The controller equation is:

$$u_k = \beta u_{k-1} + (1 - \beta)u_{k-\theta-1} + K_1(e_{k-1} - \alpha e_{k-2}) \qquad (8.17)$$

in which $K_1 = (1 - \beta)/[K(1 - \alpha)]$ (8.18)
$\alpha = \exp(-\Delta t/\tau)$ (8.19)
$\beta = \exp(-\Delta t/\tau_R)$ (8.20)
$e = y - y_{sp}$, the deviation of the process output from setpoint
Δt = sampling time
K = gain process model
τ = lag of the process model
θ = process dead time/sampling time
τ_R = lag time for desired closed loop response
u_k = process input (control signal)

This controller performs well, especially for first order and dead time systems. The major disadvantage, however, is that it does not allow the user to tune the process model on-line by comparing model output with actual process output. Also, the value of the control input $\theta/\Delta t$ time slots ago still has to be stored each execution interval. A starting value of $\tau_R = \tau$ offers a robust controller provided the process model is fairly accurate.

SUMMARY AND PRACTICAL SUGGESTIONS

Although these dead time compensation techniques are relatively simple to understand and implement, they have nevertheless proven themselves very effective in the process industry. The question of when a particular technique should be used over the other is basically a matter of personal preference. Table 8.1 is a comparison list for the Dynamic Reconciliator and the Smith Predictor.

Some practical guidelines for implementation are listed below:

1. Before attempting to close the supervisory feedback loop, ensure that the predicted output value (c_k or $y^*_{t=k}$) is a good fit of the actual output. This can best be done by comparing the trends or plots of the actual and the estimated values over the same time period. The same approach for updating the model parameters should be taken if process conditions have changed significantly.

2. If an analyzer is used in the feedback path, there is the possibility of unknowingly adding a dead time that will be less than the analyzer update time (the time it takes for a new reading of the same component to be determined) to the process. This depends on when the application was started. Ideally, the control action should be taken immediately after the analyzer is updated. This condition can be approached by forcing a controller tag to execute at a much faster rate (about four times as fast) than the update time, but outputting *only* on the execution following an update.

Table 8.1. A Comparison of the Dynamic Reconciliator and the Smith Predictor

Dynamic Reconciliator	Smith Predictor
Aside from the lead-lag block, which is often omitted from the implementation, no tuning constants are required	PID (Proportional-Integral-Derivative tunings are required)
Works well with a poorly tuned basic (regulatory) controller	Control performance is adequate with a poorly tuned basic (regulatory) controller
Supervisory setpoint changes can 'bump' the process, especially if the process gain is small ($< <1$). This can be avoided by various means, e.g., rate of change clamp to the setpoint of the basic controller, "filtering" the supervisory setpoint changes via a first order lag, etc.	Supervisory setpoint changes are less likely to "bump" the process
Requires parameter update when process conditions change significantly	Requires parameter update when process conditions change significantly

3. For reasons similar to those presented above, test runs to identify process models should be started immediately after the pertinent analyzer reading has been updated.

4. Tune a Smith Predictor PID controller as if no dead time exists in the process.

EXAMPLES

A dead time compensator is used in control of a process with dead time; the process input is controlled by a secondary control loop. The process is a distillation tower, the controlled variable is the concentration, and the control variable is the temperature. The simulation using the Smith Predictor is stored in worksheet SP11; the simulation using Dynamic Reconciliation is stored in worksheet DR11. Both primary and secondary controllers in the Smith Predictor strategy are PI controllers. The control structure is shown in Figure 8.5.

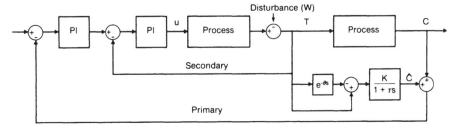

Figure 8.5. Smith Predictor control for simulation example.

Process Description

The process in the secondary loop uses the following data:

$$K = 1.0, \qquad \tau = 4 \text{ min.} \qquad \text{and} \qquad \theta = 1 \text{ min.}$$

Using a control tag processing frequency of one minute, KLAG becomes equal to $4/(4 + 1) = 0.8$.

The process equation becomes:

$$T_k = 0.8\, T_{k-1} + 0.2\, u_{k-2} * 400 + w_{k-1} \tag{8.21}$$

The constant 400 is introduced to size controller output between 0 and 1, which could correspond to 0 and 100% valve opening.

The process in the primary loop is characterized by the following data:

$$K = 0.02, \qquad \tau = 10 \text{ min.} \qquad \text{and} \qquad \theta = 13 \text{ min.}$$

The process equation is:

$$c_k = 0.91(c_{k-1} - 4) + 0.09 * 0.02(T_{k-14} - 200) + 4 \tag{8.22}$$

Controller

The secondary loop uses a PI controller. Good controller tuning parameters are a gain K_c equals 3.0 and the integral time τ_i equals 10 minutes. The error is defined as:

$$e_k = SP_k - T_k \tag{8.23}$$

and the controller equation as:

$$u_k = u_{k-1} + \frac{K_c}{400} \left[e_k - e_{k-1} + \frac{1}{\tau_i} e_k \right] \tag{8.24}$$

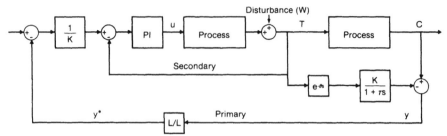

Figure 8.6. Dynamic reconciliation applied in simulation example.

The division by 400 brings u_k in the range $0 \ldots 1$ (corresponding to 0 to 100% valve opening).

The primary loop uses a PI controller; the algorithm is similar to equation (8.24).

Smith Predictor

The Smith Predictor uses the process model and the filtered process input using the following equation:

$$c_k = 0.91\, c_{k-1} + 0.09 * 0.02\, (T_{k-1} - T_{k-14}) \qquad (8.25)$$

The control strategy using dynamic reconciliation is shown in Figure 8.6. The primary and secondary process are the same as for the Smith Predictor. Also, the secondary controller is the same as before. The primary "controller" is $1/K = 1/0.02$.

The equation of the lead-lag element in the feedback path is:

$$y_k^* = 0.8 y^*_{k-1} + 0.2 y_k + 0.8(y_k - y_{k-1}) \qquad (8.26)$$

where the coefficient 0.8 determines the amount of lead action. The lead action is adapted to the disturbance pattern: for slow changes the lead should be increased; for step changes the lead action is not required at all.

CHAPTER 9

Inferential Control and Model Identification

INFERENTIAL CONTROL

In inferential control, a control scheme uses an inferential variable as the controller PV. The inferential variable can be either measured or calculated and usually infers another variable which cannot be measured or is difficult to measure. A well-known example is found in distillation columns where temperature changes reflect composition changes. If a reliable analyzer is not available, it may be possible to use the temperature to infer the composition. But even if the analyzer is available, temperature changes can often be detected well before composition changes, and, therefore, it offers an advantage to include the temperature measurement in a control scheme. Another advantage of using the temperature (inferred concentration) measurement is that, if the analyzer is temporarily out of service, control can still continue using the inferred concentration.

Inferential variables do not have to be used only for feedback control — they can be used in feed-forward control and multivariable control as well. If, for example, the stripping section temperature of a distillation column increases (as an indication of changing bottom composition), this temperature could be used as a feed forward to an overhead controller that adjusts the reflux in order to avoid a change in overhead composition. An inferential variable does

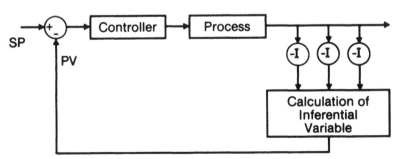

Figure 9.1. Inferential control scheme.

not necessarily have to be calculated from one measurement — more measurements could be involved. Figure 9.1 shows the concept of inferential control.

The controller structure could be simple PI control algorithm, or, in case of a major dead time, a dead time compensator as discussed in the previous chapter could be used. Table 9.1 gives some examples of inferential control. It should be mentioned that selection of inferential variables is not always obvious (it may require a good process understanding), and often extensive data collection and testing is required to find a unique relationship between the inferred and actual variable.

REASONS FOR INFERENTIAL CONTROL

The main reason for controlling an inferential variable is that the variable that should be controlled cannot be measured directly or there is not sufficient economic justification for its measurement. Analyzers, for example, are still relatively expensive and require a considerable amount of maintenance, which may force us to infer the required measurement. Also process conditions, e.g., high temperature and high pressure, may be such that actual process variable measurement is difficult and it is easier to infer the measurement.

In many light ends distillation towers, temperature can be correlated to concentration by means of a simple relationship:

$$c = -a_1 T + b_1 \tag{9.1}$$

in which c = component concentration
T = temperature
a_1, b_1 = constants

Table 9.1. Examples of Inferential Control

Variable to be Inferred	Inferential Measurement
distillation tower composition	tray temperature(s), pressure and flow(s)
flooding in distillation towers	differential pressure, tower overhead vapor flow, calculation
internal liquid flow in distillation tower	internal reflux rate
distillation tower pressure	reboiler and condensor duties
reaction rate	reactor temperature difference
fuel gas heating value	specific gravity
Mooney	viscosity
heat exchanger fouling	overall heat transfer coefficient
tank level	flow rates
composition in in-line blending	composition and flow rates of individual components

In high purity towers often a logarithmic relationship is found:

$$\log c = -a_2 T + b_2 \tag{9.2}$$

It may also be possible to correlate product composition to external tower variables such as:

$$\log c = \frac{a_3 Q_h - a_4 IR}{F} + b_3 \tag{9.3}$$

in which $\dfrac{Q_h}{F}$ = tower heat input per unit feed

$\dfrac{IR}{F}$ = internal reflux per unit feed

If a relationship such as equation (9.1) is found, the tray temperature could be used as an inner loop of a cascade control structure that eliminates disturbances quickly because the temperature responds to disturbances much faster than the analyzer does. The analyzer could then be used as the outer loop of the cascade to reset the setpoint of the inner loop. This is illustrated in Figure 9.2.

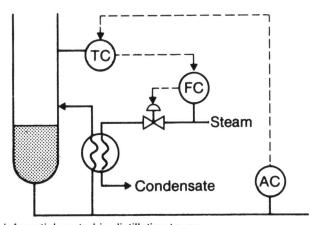

Figure 9.2.   Inferential control in distillation tower.

If a relationship such as equation (9.2) is valid in a certain situation, then control of temperature (as inferred composition) is even more useful and could lead to better control performance. In a basic control system, a logarithmic function may not be available. If the straight analyzer reading would then be used for control, we would have to control a system with a variable process gain. From equation (9.2) this gain can be derived:

$$K_p = \frac{dc}{dT} = -a_2 e^{-a_2 T + b_2} = -a_2 c \qquad (9.4)$$

When a disturbance comes in, resulting in an increase in temperature, the process gain decreases. It is known that a process with a variable gain is difficult to control by using simple PID control. If the temperature was controlled, however, disturbances would mostly be eliminated by the temperature controller. Straight analyzer to temperature control could function satisfactorily since temperature and concentration changes would be restricted to a limited operating range, and the process gain would vary only slightly.

INFERENTIAL VARIABLE SELECTION

As already mentioned, the selection of an inferential variable for control is not always obvious. In some cases, however, a process model consisting of energy, mass, and partial mass balances may give an indication which variable(s) to select. Consider a simple example. Assume an ideally mixed adiabatic reactor for which the following static partial mass balance may be written:

$$F (c_{in} - c_{out}) = Vr \qquad (9.5)$$

with   V = reactor volume
       c = reactant concentration
       F = volumetric flow
       r = reaction rate

and the static energy balance:

$$F\rho Cp (T_{in} - T_{out}) = -Vr\Delta H \qquad (9.6)$$

with ρ = density
       c_p = specific heat
       T = temperature
       ΔH = heat of reaction, negative for exothermal reactions

Combining equations (9.5) and (9.6) and using the definition for conversion,

$$C = 1 - \frac{c_{out}}{c_{in}} \tag{9.7}$$

it can be shown that:

$$C = \frac{-\rho c_p}{c_{in}\Delta H}(T_{in} - T_{out}) \tag{9.8}$$

If the reactor inlet concentration was known and it was possible to measure and calculate the other variables, the reactor conversion could be inferred from the above equation. Although in reality models may be more complex, it may be possible reduce model complexity in a certain operating region.

Another important source for inferential variable selection is past experience. Past experience often has no physical background but was obtained by monitoring and analyzing the process. The role of operators in this area should not be underestimated. An empirical model can be just as good as a model based on mass and energy balances.

DYNAMIC CONSIDERATIONS

In this section it will be illustrated, by means of a simple example, that the selection of an inferential measurement should be based not only on static considerations but also on dynamics. Equation (9.8) showed a unique relationship between conversion and temperature difference. Assume that the temperature difference has to be controlled by manipulating the inlet temperature. It can be seen from equation (9.8) that maintaining a constant $\Delta T = (T_{out} - T_{in})$ results in a constant conversion. Although this is statically true, dynamically there are some problems. When the inlet temperature is changed stepwise, the outlet temperature will lag behind as shown in Figure 9.3.

Although in Figure 9.3 it was assumed that a one degree change in inlet temperature results in a one degree change in outlet temperature, this is usually not the case in chemical reactors due to the reaction that takes place. If the relationship between a change in inlet and a change in outlet temperature is, for example:

$$\delta T_{out} = 1.2 \frac{e^{-3s}}{1 + 2s} \delta T_{in} \tag{9.9}$$

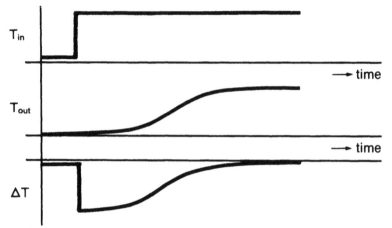

Figure 9.3.   Temperature response of chemical reactor.

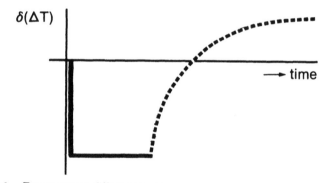

Figure 9.4.   Temperature difference response.

then the change in temperature difference is:

$$\delta(\Delta T) = \left[1.2\frac{e^{-3s}}{1 + 2s} - 1 \right] \delta T_{in} \qquad (9.10)$$

The response is shown in Figure 9.4.

Since this is a non-minimum phase response, control may be difficult. Model-based control would have to be used in order to make tight control possible. A control scheme that uses this concept is shown in Figure 9.5.

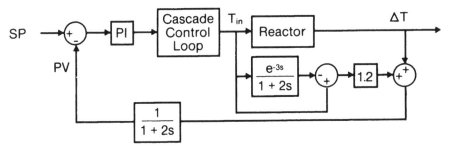

Figure 9.5. Model-based control to eliminate non-minimum phase behavior.

The PV can be written as:

$$PV = \left[\frac{1.2 \ e^{-3s}}{1 + 2s} \, \delta T_{in} - \delta T_{in} + 1.2 \ \delta T_{in} - \frac{1.2 \ e^{-3s}}{1 + 2s} \, T_{in} \right] \frac{1}{1 + 2s} \qquad (9.11)$$

$$= \frac{0.2 \delta T_{in}}{1 + 2s}$$

In other words, the PV indicates the future value of the temperature difference via a first order lag, hence it can now be used to predict final reactor conversion. When considering inferential measurements, dynamic considerations should not be ignored.

EXAMPLES

In this section some examples of inferential control will be discussed in more detail. The first example in Table 9.1 was distillation tower composition control by controlling tray temperature. In binary columns at constant pressure there is a unique relationship between temperature and composition. Although most industrial columns are not binary, it is often still possible to select a tray temperature that correlates fairly well with a key composition.

In an effort to minimize energy consumption in distillation towers, many control schemes use pressure minimization. In case of upsets, however, pressure can vary considerably. In that case it is advised to compensate the temperature measurement for changes in pressure:

$$T = T_{actual} + \frac{dT}{dP} (P_{ref} - P_{actual}) \qquad (9.12)$$

in which dT/dP = the inverse of the slope of the vapor pressure curve in a reference point

P_{ref} = a reference pressure e.g. an average operating pressure
T = the temperature

dT/dP can be determined in actual plant operation by selecting two pressures close to the normal operating pressure and to determine the corresponding temperatures. It is obvious that the selection of the tray location is of ultimate importance in correlating temperature to concentration.

In order to estimate flooding in distillation towers, the differential pressure is often used as an indicator. Once the pressure difference has reached a certain level, however, flooding has usually started and vapor and/or liquid flow have to be decreased considerably in order to return to normal operation. To make the pressure drop indicator more sensitive, it would be better to measure pressure drop above and below the feed tray. Flooding usually starts either above or below the feed tray. When the upper part starts flooding, the pressure drop in the lower part will decrease even initially, and, thus, we could detect flooding earlier than when there was only one Δp measurement available.

Another indicator for flooding is a tower mass balance. In a flooding situation more mass is entering than leaving the tower. Also, here the problem is that the tower is usually flooded before it is detected. An indicator that was found to be very useful is the tower overhead vapor flow, or the sum of reflux and distillate flow. A maximum flow limit has to be determined empirically. An advanced control strategy could take appropriate action (e.g., reduce the reflux flow) once a predetermined maximum value is exceeded.

Internal liquid flow in distillation towers can be controlled by manipulating internal reflux to the tower. Internal reflux control will minimize changes in fractionation due to condenser disturbances. It is often applied in towers with flooded condensers where the reflux can be considerably subcooled. The degree of subcooling will affect the internal liquid flow in the tower. The expression for the internal reflux flow can be derived from an energy balance around the top of the tower: internal reflux is equal to the external reflux plus the amount of overhead vapor condensed by the subcooled reflux and therefore:

$$ER = \frac{IR}{1 + \dfrac{c_p(T_{OH} - T_R)}{\Delta H}} \qquad (9.13)$$

in which   IR = interal relux
ER = external reflux from reflux drum to tower
c_p = liquid specific heat
ΔH = heat of vaporization
T_{OH} = overhead vapor temperature
T_R = reflux liquid temperature

In a cascade control scheme, the internal reflux setpoint IR_{SP} could be manipulated by an analyzer controller. Two things should be pointed out when applying internal reflux control. First, the overhead vapor temperature should always be greater than the reflux liquid temperature. An application should verify this condition before calculating equation (9.13). Second, there may be instances where the internal reflux controller may show an inverse response. If the heavy components in the feed increase, T_{OH} will increase before the reflux temperature changes, and the reflux will initially decrease. This, however, is not desirable since we would like the reflux to increase. The analyzer controller will increase the reflux once more heavies are detected in the overhead. To get around the problem either an average but constant T_{OH} could be used or a feed-forward strategy using a feed analyzer should be implemented.

Control of distillation tower pressure is usually done by feedback control to the heat input (e.g., steam) or condenser duty (e.g., cooling water flow). Figure 9.6, however, shows a situation where pressure control would benefit from feed-forward of reboiler and condenser duties. The pressure controller of tower one would be helped considerably if the heat input to tower two and the condenser duty of tower one are used in a feed-forward loop. The pressure response of tower one is:

$$\delta P = G_1 \delta F_{S1} + G_2 \delta F_{S2} - G_3 \delta F_{C1} \qquad (9.14)$$

in which δP = change in pressure
δF = change in flow
G = transfer function

In order for δP to be equal to zero, equation (9.14) has to be rewritten as:

$$\delta F_{S1} = \frac{G_3}{G_1} \delta F_{C1} - \frac{G_2}{G_1} \delta F_{S2} \qquad (9.15)$$

Equation (9.15) would be the model for the feed-forward compensator.

The last example of inferential measurement that was mentioned was the use of specific gravity in inferring the heat of combustion. For many hydrocarbons the heat of combustion is linearly dependent on the density. Thus, density can be used as an indicator of the heat of combustion. If a change in density is measured, the flow could be adjusted to maintain a constant heating value of the gas.

MODEL IDENTIFICATION

In the previous two chapters model-based control was discussed using simple models. The response of the overhead or bottom composition to feed,

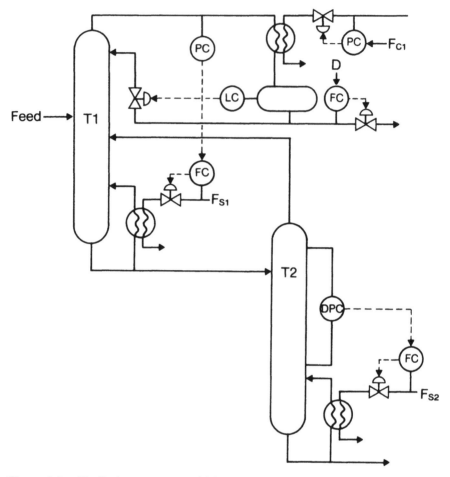

Figure 9.6.   Distillation process in which heat and condenser duties can be used in pressure control.

reflux, and reboiler changes in many distillation towers looks like an S-shaped curve. For control purposes it is often sufficient to approximate this S-shaped curve by a first order lag plus dead time model. There are many guidelines in the literature for how to make the best approximation. One that has proven to be fairly accurate is based on 25% and 75% response times. The first order time constant is determined from:

$$\tau = 0.9 \, (t_{75} - t_{25}) \tag{9.16}$$

and from the dead time θ from:

$$\theta = 0.25\ (5t_{25} - t_{75}) \tag{9.17}$$

From experimental data on a reactor it was found that t_{25} = 3.6 min and t_{75} = 5.8 min, therefore τ = 0.9 · (5.8 – 3.6) = 2 min and θ = 2.25 * (5 * 3.6 – 5.8) = 3.1 min. The actual response and approximation are shown in Figure 9.7. t_{25} is defined as the time in which the response has achieved 25% of its final value. In this case the temperature change is 142 – 130 = 12°C, hence 25% of the response is reached when the outlet temperature is equal to 130 + 0.25 * 12 = 133°C. Similarly t_{75} is defined as the time in which the response has achieved 75% of its final value, which is 139°C.

A drawback of formulas like equation (9.16) and (9.17) is that they can only be used if the process input shows a stepwise change. Often this is hard to accomplish, especially when the process input variable that has to be changed is a controlled variable itself, e.g., if the reactor inlet temperature could only be changed by changing the steam flow to a heat exchanger, the temperature inlet response itself could look like an S-shaped curve.

Another method of model identification which is often used is a PRBS (pseudo-random binary sequence) test, the data of which is fitted to a process model using a least squares technique.

A simple technique of model identification is developed here using Lotus 1-2-3; the process input may have any form or shape. The procedure is as follows:

1. Change the process input, and collect input and output data at a suitable sampling frequency:

$$\Delta t \leq 1/3\ \text{MIN}\ (\theta, \tau)$$

2. Define deviation variables; if the process input is x and the output y, define x – x_{ss} and y – y_{ss} where x_{ss} and y_{ss} are the steady state values.

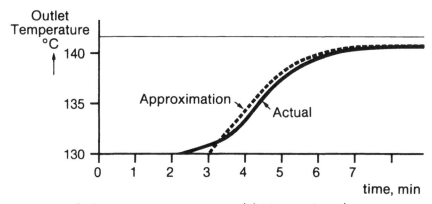

Figure 9.7. Outlet temperature response to inlet temperature change.

3. Assume a dead time θ.
4. Delay $(x - x_{ss})$ by $\theta/\Delta t$ time intervals and call the resulting variable:

$$xd_k = (x - x_{ss})_{k-\theta/\Delta t-1} \qquad (9.18)$$

5. Assume a first order time constant and process gain K and filter the delayed variable xd; call the resulting variable:

$$xdf = KLAG* xdf_{k-1} + K* (1-KLAG)* xd_{k-1} \qquad (9.19)$$

with $KLAG = \tau/(\tau + \Delta t)$
Δt = sampling interval

6. Compare xdf with $(y - y_{ss})$ and adjust the parameters if necessary.

A typical response is shown in Figure 9.8.
Parameter adjustment (K, τ, θ) should be made in the following order:

1. Adjust the process gain until the tail of the actual response and the approximation match. In this example the gain is too small.
2. Adjust KLAG until the slope of the actual response and the approximated response are the same. In this example the time constant was too large, and, therefore, KLAG is too large.
3. Adjust the dead time. If the dotted curve still starts too late, decrease the dead time.

It is advised to use the same sampling interval Δt during identification and actual control. In that case it is not necessary to calculate τ, but KLAG can be used directly.
This technique is now applied to the data in Table 9.2, which shows temperature

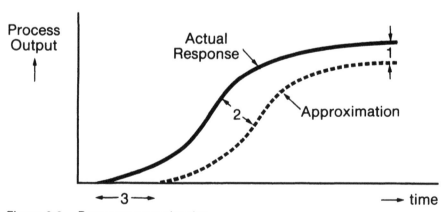

Figure 9.8.   Response approximation.

Table 9.2.   Temperature - Analyzer Data for a Light Ends Tower

A time	B temp	C conc	D delta T	E delta c	F del Td	G del Tdf
0.1	205.60	2.65				
0.2	205.40	2.73				
0.3	205.80	3.14				
0.4	205.80	3.35				
0.5	205.50	3.87				
0.6	202.70	3.82				
0.7	199.30	3.72				
0.8	196.90	4.74				
0.9	195.90	5.80				
1.0	195.70	9.19				
1.1	195.80	11.23				
1.2	195.80	12.58				
1.3	196.00	13.26				
1.4	195.90	13.47				
1.5	195.90	13.53				
1.6	195.80	13.37				
1.7	196.00	13.37				
1.8	196.00	13.58				
1.9	196.00	13.59				
2.0	195.90	13.63				
2.1	195.90	13.64				
2.2	195.90	13.72				
2.3	196.00	13.85				
2.4	196.10	14.11				
2.5	196.00	14.26				
2.6	196.10	14.41				
2.7	195.90	14.62				
2.8	196.00	14.83				
2.9	196.00	14.89				
3.0	196.80	14.92				
3.1	200.70	14.99				
3.2	203.60	15.00				
3.3	205.50	14.39				
3.4	205.90	14.27				
3.5	206.40	8.78				
3.6	206.10	8.39				
3.7	206.30	6.35				
3.8	206.00	6.26				
3.9	206.10	5.81				
4.4	206.10	5.84				

concentration data for a light ends distillation column. Data was collected every six minutes, and temperature is used to infer concentration.

Since there is no real steady state, the first five readings in Column B and C are

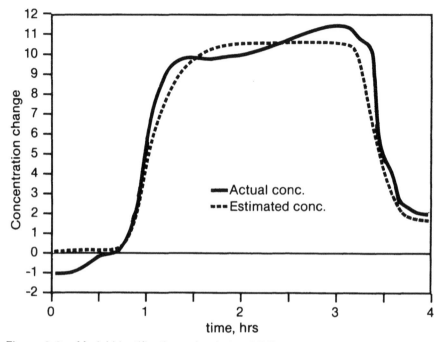

Figure 9.9.   Model identification using Lotus 1-2-3.

averaged, and the average is used as steady state value, which gives $T_{ss} = 205.5$ and $C_{ss} = 3.7$. Column D can now be constructed since D[I] = B[I] – 205.5. Similarly, column E can be constructed as E [I] = C[I] – 3.7. Column D and E are the temperature and concentration in terms of deviation variables. Column F is used to delay column D by, e.g., two time slots, thus F[4] = D[2], and so forth. In that case elements up to F[4] are made equal to zero. The first element in column G becomes zero. The next element uses equation (9.19), e.g., G[3] = 0.7* G[2] – 1.1* 0.3* F[2], in which KLAG = 0.7 and the process gain is equal to –1.1. It should be noted that a decrease in temperature leads to an increase in concentration, hence the process gain is negative.

If necessary, the values of the delay time, process gain, and lag time can now be adjusted according to the procedure described before. The calculated and actual response are shown in Figure 9.9, using a dead time $\theta = 12$ min, a lag time $\tau = 13.9$ min, and a process gain of 1.1. It can be seen that the estimated concentration approximates the actual concentration well. This analysis is found to be very useful when the process input may be disturbed from its steady state. Now a computer tag using this model can be built, and fine tuning of model parameters can be done on-line by comparing actual and calculated concentration.

CHAPTER 10

Feed-Forward Control

Feed-forward control can be usefully applied whenever a process disturbance can be measured. The general concept of feed-forward control is shown in Figure 10.1.

Assume the process output is affected by a disturbance via transfer function G_1 and by a control variable via a transfer function G_2. The disturbance is added to the control signal via a transfer function G_{FF}. Ideally the process output y does not change when the disturbance comes in, therefore the control u does not have to change. A change in y can be written as:

$$\delta y = G_1 \delta w - G_{FF} G_2 \delta w \tag{10.1}$$

which gives with the above conditions:

$$G_{FF} = \frac{G_1}{G_2} \tag{10.2}$$

Not in all cases can a useful expression for G_{FF} be derived. If, for example, the path G_1 is faster than the path G_2, feed-forward action always comes too late. This can be easily seen from the following example:

$$G_1 = \frac{e^{-3s}}{1 + 30s} \tag{10.3}$$

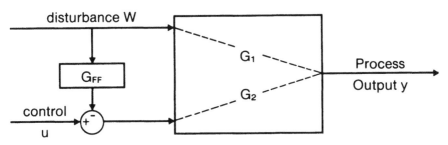

Figure 10.1. Concept of feed-forward control.

$$G_2 = \frac{e^{-5s}}{1 + 30s} \tag{10.4}$$

then:

$$\frac{G_1}{G_2} = e^{+2s} \tag{10.5}$$

which cannot be realized. In this case the best thing that can be done is to use a lead-lag for G_{FF} with the lead larger than the lag. This would give an initial kick to the control variable. Feed-forward control in a system without a major time delay is usually simple to implement; however, if time delays are present and are compensated for in the feedback loop, then an added degree of complexity is introduced, making feed-forward control more difficult to implement. In this chapter the simplest form of feed-forward control will be discussed and gradually developed for more complex systems.

RATIO CONTROL

Ratio control is a form of feed-forward control found in, for example, blending systems. No dynamics are involved, and the feed-forward element G_{FF} does not have any dynamics. An example is shown in Figure 10.2. The only part of G_{FF} that is active is a gain. Flow F_1 is multiplied by a ratio R and RF_1 is the setpoint of the flow controller for flow F_2. The ratio R can either be operator entered or be reset by a feedback control loop on the blended flow (e.g., analyzer).

Ratio control where dynamic compensation (G_{FF}) is required is found in reflux to feed ratio control in a distillation tower, as shown in Figure 10.3. Both the feed F and the reflux R affect the temperature in the tower (inferred composition). In this case the feed is the disturbance variable and the reflux the control variable, hence:

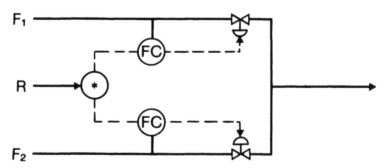

Figure 10.2.   Ratio control in a blending system.

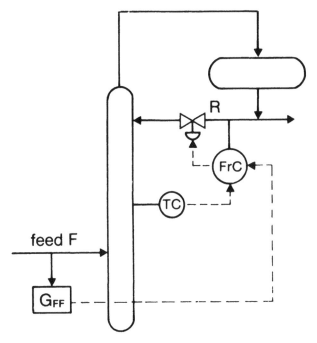

Figure 10.3.   Reflux to feed ratio control.

$$\delta T = G_1 \delta F - G_{FF} G_2 \delta R \qquad (10.6)$$

Usually reflux changes act faster than feed changes, giving G_{FF} the form of a lead-lag with a small dead time. The temperature controller can be used to reset the ratio R/F in case feed-forward action alone is not sufficient (e.g., due to other disturbances or modeling errors).

Note: If ratio control is used, the gain of G_{FF} should be made equal to one. The gain will be introduced into the system via the value of the ratio.

TRUE FEED-FORWARD CONTROL

A simple example of true feed-forward control can again be found in a distillation tower (Figure 10.4). An analyzer controller is manipulating a relatively small distillate flow. If only the level in the reflux drum was manipulating the reflux flow, and if the reflux drum was also large, then the reflux would be adjusted extremely slowly, resulting in poor analyzer response. The response, however, can be improved by adding feed-forward action of the distillate flow to the reflux flow. As soon as the distillate flow changes, the

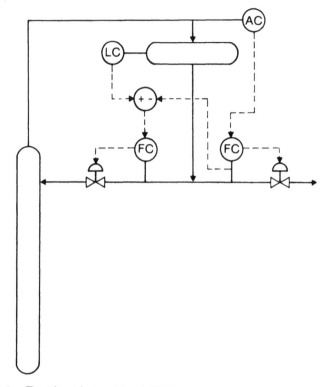

Figure 10.4.   True feed-forward in distillation.

reflux is adjusted, hence the time lag of the reflux drum will be virtually eliminated. Feed-forward control is used in a true form, since any increase or decrease in the distillate flow is passed on to the reflux flow.

FEED FORWARD IN A SMITH PREDICTOR

So far feed-forward control was relatively simple and assumed that no major dead time was involved. If, however, the process has a major dead time and model-based control is applied (e.g., a Smith Predictor) feed forward becomes somewhat more difficult to implement. The implementation is easier to understand if a specific example is given. Assume that it was not possible to find a temperature that could be used as an inferential measurement for the overhead impurity in a distillation tower. Test runs have been carried out and the following transfer function is found from reflux to analyzer reading:

$$\frac{\delta A}{\delta R} = \frac{-0.03e^{-18s}}{1 + 19s} \tag{10.7}$$

in which δA = change in analyzer reading
            δR = change in reflux

Because of the relatively large time delay, we decide to use a Smith Predictor as a dead time compensator. The control tag will run every two minutes. Using equations (7.27) and (7.29) this equation may be written in z-transform:

$$\frac{\delta A}{\delta R} = \frac{-0.003z^{-10}}{1-0.90z^{-1}} \qquad (10.8)$$

The Smith Predictor control structure is shown in Figure 10.5.

The controller gain is shown as K_C; a is a function of the integral time τ_i:

$$a = 1+\frac{2}{\tau_i} \qquad (10.9)$$

It was found that control worked well, except for large feed changes. Therefore, it was decided to use feed-forward control in order to improve control performance further. From another test run the transfer function from feed to analyzer was found:

$$\frac{\delta A}{\delta F} = \frac{0.015e^{-20s}}{1 + 13.3s} \qquad (10.10)$$

or in z-notation

$$\frac{\delta A}{\delta F} = \frac{0.0021z^{-11}}{1-0.86z^{-1}} \qquad (10.11)$$

It is clear that the feed-forward action should be added between the output of the controller and the input to the process. Incremental feed forward will be

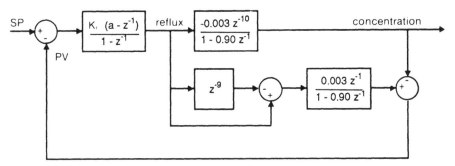

Figure 10.5.   Analyzer control using Smith Predictor.

used rather that absolute feed forward since in a steady state situation feed forward should not contribute to control. The purpose of feed forward is to have such an impact on the process that the analyzer reading does not change, and, therefore, the feedback control loop does not have to act. For that to be the case the model part in Figure 10.5 should not be affected by feed-forward control and the addition of feed forward is therefore just before the process but after the junction where the reflux is fed into the Smith Predictor. The feed also has to be dynamically compensated. G_{FF} can easily be calculated from Figure 10.6.

$$\frac{0.0021z^{-11}}{1-0.86z^{-1}} = G_{FF}\frac{0.003z^{-10}}{1-0.90z^{-1}} \tag{10.12}$$

or

$$G_{FF} = 0.7z^{-1}\frac{1-0.90z^{-1}}{1-0.86z^{-1}} \tag{10.13}$$

It can be seen that equation (10.13) represents a one-cycle delay and lead-lag function. The lead-lag part can be written as:

$$\frac{y}{x} = 0.7\frac{1-0.90z^{-1}}{1-0.86z^{-1}} \tag{10.14}$$

or

$$\begin{aligned} y_k &= 0.86\ y_{k-1} + 0.7x_k - 0.7*0.90*\ x_{k-1}\\ &= 0.86\ y_{k-1} + 0.5*0.14x_k + 0.63\ (x_k - x_{k-1}) \end{aligned} \tag{10.15}$$

Let us look again at equations (10.7) and (10.10). Let $K_1 = 0.03$, $\theta_1 = 18$, $\tau_1 = 19$, $K_2 = 0.015$, $\theta_2 = 20$, and $\tau_2 = 13.3$. Then it can be seen that (compare equation [7.22]):

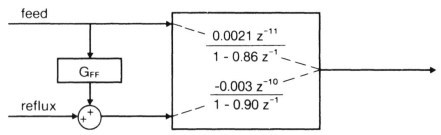

Figure 10.6. Determination of feed-forward action.

$$KLAG = e^{-\Delta t/\tau_2} = e^{-2/13.3} = 0.86$$

$$\frac{K_2}{K_1} = \frac{0.015}{0.03} = 0.5$$

$$1\text{-}KLAG = 1\text{-}e^{-\Delta t/\tau_2} = 0.14 \qquad (10.16)$$

$$KLEAD = \frac{K\tau_1}{\tau_2 + \Delta t} = \frac{0.5* 19}{15.3} = 0.63$$

The control scheme with feed-forward action can now be constructed and is shown in Figure 10.7.

The scheme may need some explanation. The major reason to use an incremental feed forward is that in a steady state situation the feed-forward control signal does not contribute to the process input:

$$\Delta F_k{}^* = (1\text{-}z^{-1})F_k{}^*$$
$$= F_k{}^* \text{-} F_{k-1}{}^* \qquad (10.17)$$

in which $F_k{}^*$ is the dynamically compensated feed at time k. In a steady state situation $\Delta F_k{}^* = 0$. However, if the feed F^* is multiplied by $(1\text{-}z^{-1})$, it also has

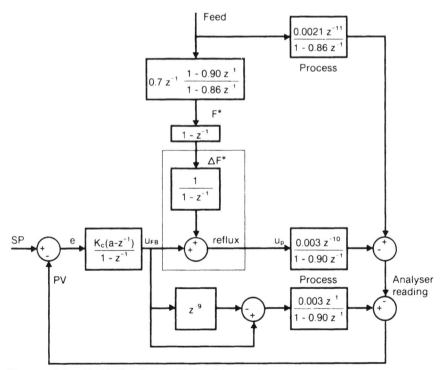

Figure 10.7. Smith Predictor with feed forward.

to be divided by it in order to leave F* unchanged and avoid adding additional dynamics. This division together with the addition of the feed forward to the feedback signal is shown in the encircled area in Figure 10.7. The following equation holds at this point:

$$\frac{\Delta F_k^*}{(1-z^{-1})} + u_{FB,k} = u_{P,k} \tag{10.18}$$

which can be written as:

$$u_{P,k} = u_{P,k-1} + (u_{FB,k} - u_{FB,k-1}) + (F_k^* - F_{k-1}^*) \tag{10.19}$$

current value	previous value	change in reflux	change in reflux
of the reflux	of the reflux	due to feedback	due to feed
			forward

This equation is used in a practical implementation. Although $u_{FB,k} = u_{SP,k}$, the input to the Smith Predictor, the practical implementation will use the following equation:

$$u_{SP,k} = u_{SP,k-1} + u_{FB,k} - u_{FB,k-1} \tag{10.20}$$

| current input to | previous input to | change in reflux |
| Smith Predictor | Smith Predictor | due to feedback |

The advantage of the use of equations (10.19) and (10.20) in any practical application is that limit checks can be made before feed-forward and feedback incremental control signals are used. This is the necessary security checking part of any advanced control application. Also this form of implementation provides a simple way of incorporating feed-forward control in feedback loops that are dynamically compensated.

Example

Let us write equation (10.13) as:

$$F_k^* = 0.86 F_{k-1}^* + 0.07F_{k-1} + 0.63(F_{k-1} - F_{k-2}) \tag{10.21}$$

in which F_k = feed at time k
F_k^* = filtered feed at time k

Assume the steady state values of the feed and reflux are 300 and 200 barrels/hour respectively. The feed shows a step increase to 320 barrels/hour. Table 10.1 shows how the reflux will change in time.

Another possibility, which is merely a reconstruction of Figure 10.7 and which is sometimes easier from an implementation point of view, is shown in

Table 10.1.   Impact of Feed-Forward Action on Reflux

k	TIME (min.)	FEED (bbl/hr)	F^*_k	$F^*_k - F^*_{k-1}$	REFLUX (bbl/hr)
0	0	300	150	0.0	200
1	2	320	150	0.0	200
2	4	320	164.0	14.0	214.0
3	6	320	163.44	-0.56	213.44
4	8	320	162.96	-0.48	212.96
5	10	320	162.54	-0.42	212.54
6	12	320	162.19	-0.35	212.19
7	14	320	161.88	-0.31	211.88
8	16	320	161.62	-0.26	211.62
9	18	320	161.39	-0.23	211.39
10	20	320	161.20	-0.19	211.20
			. . .	. . .	. . .
			. . .	. . .	. . .
			. . .	. . .	. . .
			160.00	0.0	210.00

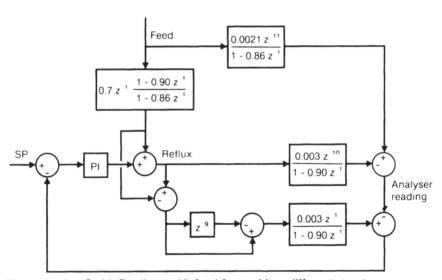

Figure 10.8.   Smith Predictor with feed forward in a different structure.

Figure 10.8. When the PI controller is in manual, the lead-lag-delay element scanning the feed flow can still provide an input to the Smith Predictor part of the control strategy.

FEED FORWARD IN DYNAMIC RECONCILIATION

In order to illustrate feed forward in a control structure that uses dynamic reconciliation as a dead time compensator, a different example will be discussed. Consider a distillation tower with two feeds, F_1 and F_2; the reflux R is used to maintain a constant overhead concentration. Since the overhead concentration A depends on reflux and feed flows, it can be written as:

$$A = A_0 - \hat{K}_1\hat{G}_1 R + \hat{K}_2\hat{G}_2 F_1 + \hat{K}_3\hat{G}_3 F_2 \tag{10.22}$$

in which $\hat{K}_1\hat{G}_1 = \delta A/\delta R$ estimated
$\hat{K}_2\hat{G}_2 = \delta A/\delta F_1$ estimated
$\hat{K}_3\hat{G}_3 = \delta A/\delta F_2$ estimated
$\hat{G}_i$ = estimated dynamic transfer function
$\hat{K}_i$ = estimated gain of the transfer function
A_0 = a bias term, is a function of steady state values

The bias term can be estimated from equation (10.22) as:

$$A_0 = A + \hat{K}_1\hat{G}_1 R - \hat{K}_2\hat{G}_2 F_1 - \hat{K}_3\hat{G}_3 F_2 \tag{10.23}$$

The control equation is derived from the static version of equation (10.22):

$$R_{SP} = \frac{A_0 - A_{SP}}{\hat{K}_1} + \frac{\hat{K}_2}{\hat{K}_1} F_1 + \frac{\hat{K}_3}{\hat{K}_1} F_2 \tag{10.24}$$

with A_{SP} = setpoint for concentration A
R_{SP} = setpoint for reflux

A control scheme that is based on equations (10.23) and (10.24) is shown in Figure 10.9. Let us now verify how good the control scheme is. A change δF_2 in feed F_2 has the following impact on concentration A:

$$K_3 G_3 \delta F_2 - \hat{K}_3 \frac{1}{\hat{K}_1} K_1 G_1 \delta F_2 = K_3 \delta F_2 [G_3 - G_1] \tag{10.25}$$

if estimated gains are equal to actual gains. Since G_3 and G_1 both have a gain of unity, statically there will be no impact of the feed change on the concentration. However, dynamically there is an impact when G_3 and G_1 are not the same (which is likely to be the case). Therefore the feed-forward paths will be modified to include dynamics in order to fully eliminate the impact of feed changes on concentration. The result is the control scheme shown in Figure 10.10 which is called extended dynamic reconciliation (EDR). Now the impact of a change on A is:

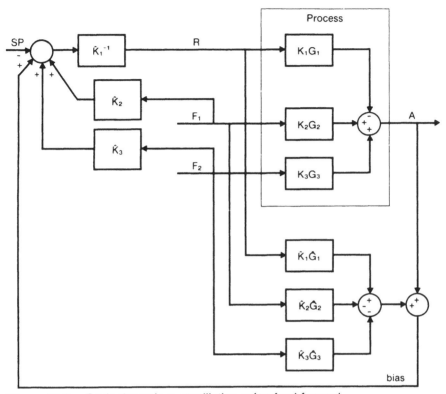

Figure 10.9. Basic dynamic reconciliation using feed forward.

$$K_3 G_3 \delta F_2 - \frac{\hat{G}_3}{\hat{G}_1} \hat{K}_3 \frac{1}{\hat{K}_1} K_1 G_1 \delta F_2 = 0 \qquad (10.26)$$

in case of no modeling error. It should be noted that all parameter values in Figure 10.10 are absolute values.

In order to illustrate the control scheme, assume the following transfer functions have been found from test runs:

$$\frac{\delta A}{\delta R} = \frac{-0.03 e^{-26s}}{1 + 16s} \qquad (10.27)$$

$$\frac{\delta A}{\delta F_1} = \frac{0.02 e^{-26s}}{1 + 20s} \qquad (10.28)$$

$$\frac{\delta A}{\delta F_2} = \frac{0.01 e^{-40s}}{1 + 4s} \qquad (10.29)$$

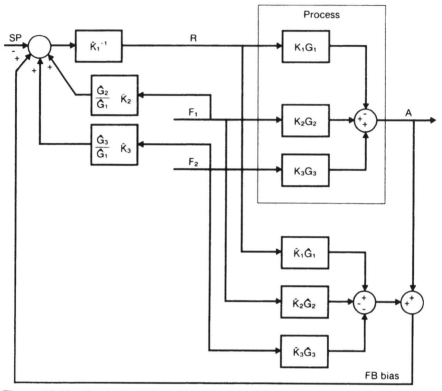

Figure 10.10. Implementation of feed forward in control scheme using extended dynamic reconciliation (EDR).

If the control tag will be executed every two minutes, the following functions can be defined:

$$\hat{K}_1 = 0.03 \tag{10.30}$$

$$\hat{G}_1 = \frac{e^{-26s}}{1 + 16s} = \frac{0.12z^{-14}}{1-0.88z^{-1}} \tag{10.31}$$

$$\hat{K}_2 = 0.02 \tag{10.32}$$

$$\hat{G}_2 = \frac{e^{-26s}}{1 + 20s} = \frac{0.10z^{-14}}{1-0.90z^{-1}} \tag{10.33}$$

$$\hat{K}_3 = 0.01 \tag{10.34}$$

$$\hat{G}_3 = \frac{e^{-26s}}{1 + 4s} = \frac{0.39z^{-21}}{1-0.61z^{-1}} \tag{10.35}$$

The dynamic element in the F_1 feed-forward path is therefore:

$$\hat{K}_2 \frac{\hat{G}_2}{\hat{G}_1} = 0.02 \cdot \frac{0.10z^{-14}}{1-0.90z^{-1}} \cdot \frac{1-0.88z^{-1}}{0.12z^{-14}}$$

$$= 0.0167 \cdot \frac{1-0.88z^{-1}}{1-0.90z^{-1}} \qquad (10.36)$$

Similarly, the element in the F_2 feed-forward path becomes:

$$\hat{K}_3 \frac{\hat{G}_3}{\hat{G}_1} = 0.01 \cdot \frac{0.39z^{-21}}{1-0.61z^{-1}} \cdot \frac{1-0.88z^{-1}}{0.12z^{-14}}$$

$$= 0.033z^{-7} \cdot \frac{1-0.88z^{-1}}{1-0.61z^{-1}} \qquad (10.37)$$

The control scheme using dynamic feed-forward models is shown in Figure 10.11. Since models are only an approximation of reality, modeling errors may exist. The best way to tune the control scheme is on-line. Without feed-forward and feedback control, change the reflux and monitor the concentration A and the feedback bias. Make sure that the time constant, time delay, and gain are estimated correctly and adjust them if necessary. The second step is to tune the feed-forward paths on-line without feedback. Keeping one feed constant (e.g., F_2), change F_1. Feed-forward control will now adjust the reflux. From the response of the concentration (and the bias) it can be concluded if the feed-forward dynamic element has to be changed or not. If a change is made, update the model from F_1 to feedback bias accordingly. Repeat the procedure for feed F_1. If a better feed-forward dynamic element in this case is, for example:

$$\frac{0.03 \, (1-0.88z^{-1})}{1-0.80z^{-1}} \qquad (10.38)$$

then the model from F_2 to feedback bias has to be changed to:

$$\frac{0.0035z^{-21}}{1-0.80z^{-1}} \qquad (10.39)$$

Usually a lead-lag element in the feedback path will not be necessary. Since feed forward will eliminate most of the disturbances, it may be advantageous to perform feedback control every time the analyzer updates and run feed-forward control fast (e.g., every two minutes).

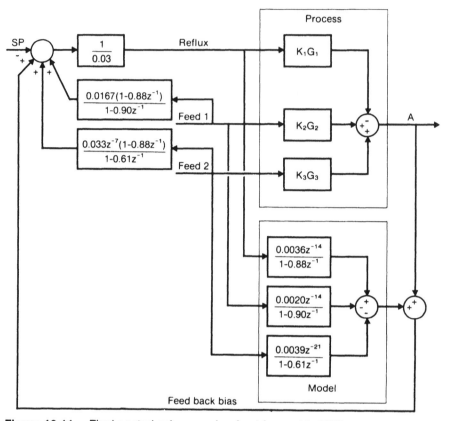

Figure 10.11.   Final control scheme using feed forward in EDR.

EXAMPLES

The diskette contains two files which illustrate how feed-forward control is implemented in a model-based control strategy: file SP11FF, illustrating feed forward in a Smith Predictor strategy, and file DR11FF, illustrating feed forward in a dynamic reconciliator control strategy. The block diagram for the Smith Predictor case is shown in Figure 10.12.

Process Description

A secondary loop is omitted (assumed to be fast) in order to fully concentrate on the effect of feed forward. The following transfer functions are given:

$$G_F = \frac{0.015e^{-20s}}{1 + 13.3s} \tag{10.40}$$

$$G_R = \frac{-0.03e^{-18s}}{1 + 19s} \tag{10.41}$$

Running a control tag every two minutes gives, for the discretized equations:

$$G_F = \frac{0.0021z^{-11}}{1 - 0.86z^{-1}} \tag{10.42}$$

$$G_R \doteq \frac{-0.003z^{-10}}{1 - 0.90z^{-1}} \tag{10.43}$$

The discrete process equations used are therefore:

$$F^*_k = 0.86\ F^*_{k-1} + 0.14*\ 0.015*\ (F_{k-11}-F_{ss}) \tag{10.44}$$

$$R^*_k = 0.90\ R^*_{k-1} + 0.1*\ 0.03*(R_{k-10}-R_{ss}) \tag{10.45}$$

with $F_{ss} = 33000$ kg/hr and $R_{ss} = 30000$ kg/hr, the steady state values for feed and reflux.

F^*_k and R^*_k are the output of transfer function G_F and G_R respectively. The analyzer reading is then:

$$A_k = A_{ss} - F^*_k - R^*_k$$

with $F_{ss} = 86$, the steady state value of the analyzer.

Controller

The feedback controller uses a simple PI control algorithm.

Feed-Forward Control Action

The feed-forward controller is $G_{FF} = G_F/G_R$

$$G_{FF} = 0.7z^{-1}\frac{1-0.90z^{-1}}{1-0.86z^{-1}} \tag{10.46}$$

or

$$F^{**}_k = 0.86\ F^{**}_{k-1} + 0.7F_{k-1}-0.7*\ 0.90\ F_{k-2}$$

$$= 0.86\ F^{**}_{k-1} + 0.5*\ 0.14F_{k-1} + 0.63\ (F_{k-1}-F_{k-2}) \tag{10.47}$$

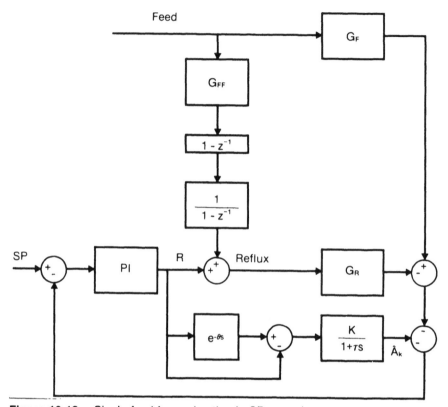

Figure 10.12. Single feed-forward action in SP control strategy.

Note: In practice the lead term (in this case 0.63) is often modified until the best response is obtained in case of feed-forward control only. This is because modeling and round-off errors have the largest impact on the lead term.

Smith Predictor

The Smith Predictor uses the G_R model and is therefore given by:

$$\hat{A}_k = 0.90\hat{A}_{k-1} + 0.10* \, 0.03* \, (R_{k-10}-R_{k-1}) \qquad (10.48)$$

with $\hat{A}_k$ the predictor output at time k.

The second simulation, DR11FF, illustrates feed-forward control in a model-based environment using dynamic reconciliation. The block diagram for this case is shown in Figure 10.13. Note the similarity and differences with Figure 10.12. The process in this case is the same as for the Smith Predictor and given by equations (10.40) and (10.41) c.q. (10.44) and (10.45). The "con-

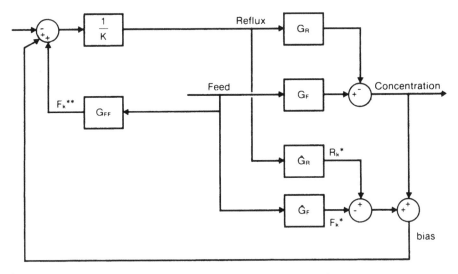

Figure 10.13.   DR with feed forward for simulation example.

troller" in this case is $1/K = 1/0.03$ with 0.03 the gain from reflux to concentration.

The feed model G_F is similar to the process model:

$$F^*_k = 0.86F^*_{k-1} + 0.14* 0.015* F_{k-11} \qquad (10.49)$$

Similarly the reflux model $\hat{G}_R$ is:

$$R^*_k = 0.90R^*_{k-1} + 0.1* 0.03* R_{k-10} \qquad (10.50)$$

The feed-forward element in this case becomes $G_{FF} = KG_F/G_R$ or:

$$G_{FF} = 0.03* 0.72z^{-2}\frac{1-0.90z^{-1}}{1-0.86z^{-1}} \qquad (10.51)$$

from which

$$F^{**}_k = 0.86F^{**}_{k-1} + 0.5* 0.14* 0.03F_{k-1} + 0.63* 0.03* (F_{k-1} - F_{k-2}) \quad (10.52)$$

CHAPTER 11

Interaction and Decoupling

Until now only systems with one manipulated variable have been discussed. The control scheme was extended with feed-forward control in case measurable disturbances were present. Feed-forward control is a form of decoupling such that changes in the disturbance variable do not affect the controlled variable. However, if the disturbance variable is another manipulated variable, decouplers is a more suitable terminology than feed forward, to indicate that control loops are decoupled and perform independently from one another.

Not in all cases is decoupling of control loops necessary; it will depend on the interaction between the control loops. In some cases it may be possible to tune certain loops much faster than others, thereby eliminating interaction to a certain extent. However, the trade-off is that control performance of the detuned control loop may be poor. It is always better to determine how much interaction there is in a particular system and use available control techniques to develop a proper control scheme. Therefore, a measure of interaction between control loops is required in order to decide whether there is any interaction or not. Another question that arises at the same time is: How do I select the proper combination of controlled and manipulated variables? To help the engineer in defining a measure of multivariable system performance, Bristol[38] developed the relative gain matrix concept.

THE RELATIVE GAIN MATRIX

Bristol[38] recommended that each open loop gain should first be evaluated with all other loops open, i.e., all other manipulated variables constant (manual mode). Determine the open loop gain of the ith controlled variable as a response to a change in the jth manipulated variable with all other manipulated variables held constant:

$$1_u = (\delta y_i / \delta u_j)u \tag{11.1}$$

in which δy_i = change in controlled variable i
δu_j = change in manipulated variable j

127

Now reevaluate the gain with all other loops closed, i.e., with all other controlled variables constant (automatic mode). The open loop gain of the ith controlled variable as a result of a change in the jth manipulated variable is then defined as:

$$l_y = (\delta y_i / \delta u_j)_y \tag{11.2}$$

An element of the relative gain matrix is now defined as:

$$\lambda_{ij} = l_u / l_y \tag{11.3}$$

and the relative gain matrix becomes:

$$\Lambda = \begin{array}{c} \\ y_1 \\ y_2 \\ : \\ y_i \\ : \end{array} \begin{array}{c} u_1 \quad u_2 \qquad\qquad u_j \\ \left[\begin{array}{cccc} \lambda_{11} & \lambda_{12} & \cdots\cdots & \lambda_{1j} & \cdots \\ \lambda_{21} & \cdots\cdots\cdots\cdots\cdots\cdots \\ & \cdots\cdots\cdots\cdots\cdots\cdots \\ \lambda_{i1} & \cdots\cdots\cdots & \lambda_{ij} & \cdots\cdots \\ & \cdots\cdots\cdots\cdots\cdots\cdots \end{array} \right] \end{array} \tag{11.4}$$

If y_i does not respond to u_j when all other manipulated variables are constant, λ_{ij} is zero and u_j should not be used to control y_i. A useful property of the relative gain matrix is that the sum of all elements in each row and each column is equal to one. If one element in a row is one and other elements in the same row all have the same sign, then these elements have to be zero which means that one manipulated variable influences only one controlled variable. In a 2×2 system this means that only one element has to be found, e.g., λ_{11}, and the others follow from:

$$\left. \begin{array}{l} \lambda_{12} = 1 - \lambda_{11} \\ \lambda_{21} = 1 - \lambda_{11} \\ \lambda_{22} = \lambda_{11} \end{array} \right\} \tag{11.5}$$

If a number in a row is greater than one, there must obviously be a negative number in the same row in order to let the sum of the elements be equal to one. Any value of λ greater than 5 indicates potential loss of control.

THE MATRIX METHOD

We will now examine a 2×2 system described by the following steady state equations:

$$y_1 = K_{11}u_1 + K_{12}u_2 \qquad (11.6)$$

$$y_2 = K_{21}u_1 + K_{22}u_2 \qquad (11.7)$$

in which

$$K_{ij} = (\delta y_i/\delta u_j)_u \qquad (11.8)$$

Equations (11.6) and (11.7) can be written as:

$$y = Ku \qquad (11.9)$$

with

$$y = \begin{pmatrix} y_1 \\ y_2 \end{pmatrix}, \quad K = \begin{pmatrix} K_{11} & K_{12} \\ K_{21} & K_{22} \end{pmatrix}, \quad u = \begin{pmatrix} u_1 \\ u_2 \end{pmatrix} \qquad (11.10)$$

However, equations (11.6) and (11.7) can also be written as:

$$u_1 = H_{11}y_1 + H_{12}y_2 \qquad (11.11)$$

$$u_2 = H_{21}y_1 + H_{22}y_2 \qquad (11.12)$$

or in matrix-vector notation:

$$u = Hy \qquad (11.13)$$

with

$$H = \begin{pmatrix} H_{11} & H_{12} \\ H_{21} & H_{22} \end{pmatrix} \qquad (11.14)$$

and

$$H_{ij} = (\delta u_j/\delta y_i)_y \qquad (11.15)$$

It can be seen that:

$$\lambda_{ij} = K_{ij} H_{ji} \qquad (11.16)$$

or in matrix notation:

$$\Lambda = KH^T \qquad (11.17)$$

Comparing equations (11.9) and (11.13) shows that

$$H = K^{-1} \tag{11.18}$$

which, with (11.17), results in:

$$\Lambda = K \ [K^{-1}]^\mathsf{T} \tag{11.19}$$

The elements K_{ij} can be obtained by putting all control loops in the process in the manual mode, changing one manipulated variable, and observing all responses in the controlled variables.

EXAMPLE

In a light ends tower the overhead composition x_D is controlled by manipulating the reflux R and the bottom composition x_B by manipulating the stripping section temperature T. In this case of a 2 × 2 system, the matrix method does not have be used since a simple expression for λ_{11} can be derived. Combination of equations (11.6) and (11.7) gives:

$$y_1 = K_{11}u_1 + \frac{K_{12}(y_2 - K_{21}u_1)}{K_{22}} \tag{11.20}$$

from which:

$$\left(\frac{\delta y_1}{\delta u_1} \right)_{y_2} = K_{11} - \frac{K_{12}K_{21}}{K_{22}} \tag{11.21}$$

From equation (11.6):

$$\left(\frac{\delta y_1}{\delta u_1} \right)_{u_2} = K_{11} \tag{11.22}$$

Therefore, the expression for λ_{11} becomes:

$$\lambda_{11} = \left(\frac{\delta y_1}{\delta u_1} \right)_{u_2} \bigg/ \left(\frac{\delta y_1}{\delta u_1} \right)_{y_2} =$$

$$\frac{1}{1 - K_{12}K_{21}/K_{11}K_{22}} \tag{11.23}$$

Assume the following static model:

$$\left. \begin{array}{l} \delta x_D = -0.03\ \delta R + 0.1\ \delta T \\ \delta x_B = +0.06\ \delta R - 1.0\ \delta T \end{array} \right\} \qquad (11.24)$$

in which δx_D = change in overhead composition
δx_B = change in bottom composition
δR = change in reflux
δT = change in stripping section temperature

Using equation (11.23) the expression for λ_{11} is:

$$\lambda_{11} = \cfrac{1}{1 - \cfrac{0.1 * 0.06}{0.03 * 1.0}} = 1.25 \qquad (11.25)$$

which gives for the relative gain matrix:

$$\Lambda = \begin{array}{c} \\ x_D \\ x_B \end{array} \begin{array}{c} R \qquad\ \ T \\ \left[\begin{array}{cc} 1.25 & -0.25 \\ -0.25 & 1.25 \end{array} \right] \end{array} \qquad (11.26)$$

which indicates moderate interaction. Ideally the diagonal elements should be equal or close to one, indicating no or little interaction. It is obvious that the most interactive 2×2 system is a system in which all elements are equal to 0.5. The relative gain matrix is a useful tool in selecting the right combination of input and output variables.[39] The input and output combination for which the relative gain is closest to one should be selected. The relative gain matrix also indicates whether decouplers are required or not. Look at the following example for a 3×3 system:

$$\Lambda = \begin{array}{c} \\ y_1 \\ y_2 \\ y_3 \end{array} \begin{array}{|ccc|} \hline u_1 & u_2 & u_3 \\ \hline 0.7 & 0.2 & 0.1 \\ -0.2 & 0.9 & 0.3 \\ 0.5 & -0.1 & 0.6 \\ \hline \end{array}$$

The obvious choice is to control y_2 with u_2; there is some interaction from the other loops. This leaves u_1 to control y_1 and u_3 to control y_3. Interaction from the first to the third loop in the system will be pronounced.

DYNAMIC RELATIVE GAIN

So far, the relative gain was calculated using process gains only. No dynamics were involved although it is relatively simple to include process dynamics. For the 2 × 2 system the relative gain λ_{11} would be, in analogy with equation (11.23):

$$\lambda_{11} = \frac{1}{1 - K_{12}G_{12}K_{21}G_{21}/K_{11}G_{11}K_{22}G_{22}} \qquad (11.27)$$

in which G equals dynamic transfer function.

Relative gain λ_{11} will now become a dynamic function; it may be difficult, however, to predict its behavior in time, depending on the various expressions for G. One simplification that can be made is when $K_{11}K_{22} >> K_{12}K_{21}$, in which case λ_{11} approaches one, indicating little or no interaction. Although dynamic analysis might be required, especially if there are strongly different dynamics, in most cases the steady state relative gain analysis is a sufficient indicator for control loop interaction. If transfer functions are the same, say G_{11}, equation (11.27) reduces to equation (11.23). In that case equation (11.21) becomes:

$$\left(\frac{\delta y_1}{\delta u_1}\right)_{y_2} = K_{11}G_{11}\left[1 - \frac{K_{12}K_{21}}{K_{11}K_{22}}\right] = \frac{K_{11}G_{11}}{\lambda_{11}} \qquad (11.28)$$

It can be seen that as the value of λ_{11} increases, y_1 becomes more insensitive to changes in u_1, and control will eventually become very ineffective.

DECOUPLING IN A SMITH PREDICTOR

Decoupler design is very much the same as design of feed forward. Consider a 2 × 2 distillation system in which overhead and bottom composition are controlled by reflux and stripping section temperature. The following transfer functions were found:

$$\frac{\delta x_D}{\delta R} = \frac{-0.03\ e^{-34s}}{1 + 17s} \qquad (11.29)$$

$$\frac{\delta x_B}{\delta R} = \frac{0.06\ e^{-30s}}{1 + 49s} \qquad (11.30)$$

$$\frac{\delta x_D}{\delta T} = \frac{0.1\ e^{-38s}}{1 + 14s} \qquad (11.31)$$

$$\frac{\delta x_B}{\delta T} = \frac{-1.0\ e^{-14s}}{1 + 14s} \qquad (11.32)$$

If the control tags for top and bottom composition control will run every two minutes, the z-transform functions become:

$$\frac{\delta x_D}{\delta R} = \frac{-0.00333 \ z^{-18}}{1 - 0.89 \ z^{-1}} \tag{11.33}$$

$$\frac{\delta x_B}{\delta R} = \frac{0.0024 \ z^{-16}}{1 - 0.96 \ z^{-1}} \tag{11.34}$$

$$\frac{\delta x_D}{\delta T} = \frac{0.0133 \ z^{-20}}{1 - 0.87 \ z^{-1}} \tag{11.35}$$

$$\frac{\delta x_B}{\delta T} = \frac{-0.133 \ z^{-8}}{1 - 0.87 \ z^{-1}} \tag{11.36}$$

The decouplers can easily be determined from Figure 11.1.
The reflux has the following impact on x_B:

$$[-G_{D1}K_{22}G_{22} + K_{21}G_{21}] \ \delta R = 0 \tag{11.37}$$

from which

$$G_{D1} = \frac{K_{21}G_{21}}{K_{22}G_{22}} \tag{11.38}$$

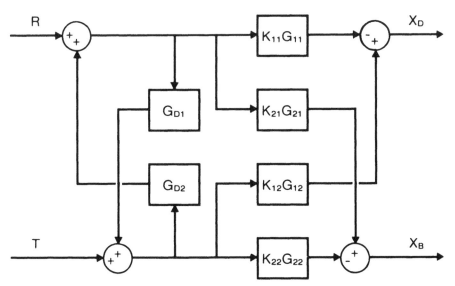

Figure 11.1. Decouplers in a 2 × 2 system.

Similarly:

$$G_{D2} = \frac{K_{12}G_{12}}{K_{11}G_{11}} \qquad (11.39)$$

Using equations (11.33) to (11.36), equations (11.38) and (11.39) become:

$$G_{D1} = \frac{0.0024 \ z^{-16}}{1 - 0.96 \ z^{-1}} \cdot \frac{1 - 0.87 \ z^{-1}}{0.133 \ z^{-8}}$$

$$\qquad (11.40)$$

$$= 0.0185 \ z^{-8} \frac{1 - 0.87 \ z^{-1}}{1 - 0.96 \ z^{-1}}$$

and

$$G_{D2} = \frac{0.0133 \ z^{-20}}{1 - 0.87 \ z^{-1}} \cdot \frac{1 - 0.89 \ z^{-1}}{0.00333 \ z^{-18}}$$

$$\qquad (11.41)$$

$$= 4.0 \ z^{-2} \frac{1 - 0.89 \ z^{-1}}{1 - 0.87 \ z^{-1}}$$

The complete block diagram for the controlled system is given in Figure 11.2; the simulation of this example is given in file SP22.

When there is no major dead time in the system the two Smith Predictors can be omitted, in which case the control system is simplified considerably. If the output from the decouplers is R* and T* respectively, the incremental signal should be calculated:

$$\Delta R^*_k = R^*_k - R^*_{k-1} \qquad (11.42)$$

and

$$\Delta T^*_k = T^*_k - T^*_{k-1} \qquad (11.43)$$

The new process input is then calculated according to:

$$T_k = T_{k-1} + \Delta u_{2,k} + \Delta R^*_k \qquad (11.44)$$

and

$$R_k = R_{k-1} + \Delta u_{1,k} + \Delta T^*_k \qquad (11.45)$$

where Δu_1 and Δu_2 are the changes in output of controller 1 (reflux) and 2 (temperature) respectively, i.e., $\Delta u_k = u_k - u_{k-1}$. The foregoing four equations

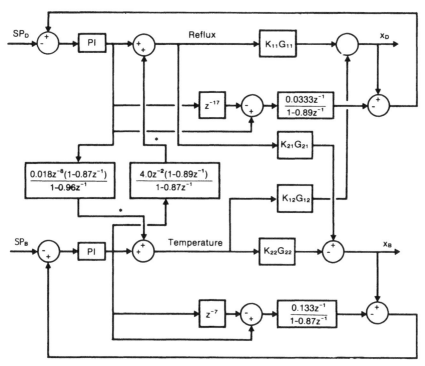

Figure 11.2. Fully decoupled 2 × 2 system.

are not shown in Figure 11.2 but are denoted by the asterisk at the decoupler outlet. This procedure is similar to the one followed in feed-forward control (see previous chapter).

It should be noted that the signal to the decouplers should be picked up at the outlet of each controller. If the decoupler signal was picked up at the process inlet, the situation should be evaluated carefully. A change δT in temperature would give a change $G_{D2} \, \delta T$ in reflux which in turn would give a change $G_{D1} \, \delta R$ in temperature. The change in temperature would therefore be:

$$\frac{\delta T_p}{\delta T_i} = \frac{1}{1 - G_{D1}G_{D2}} \tag{11.46}$$

which can easily be derived from Figure 11.3.

Equation (11.46) would become unstable when

$$G_{D1}G_{D2} = 1 \tag{11.47}$$

As long as $G_{D1}G_{D2} < < 1$ there will not be much impact. However, an easy way around the problem is the configuration as shown in Figure 11.2 where the

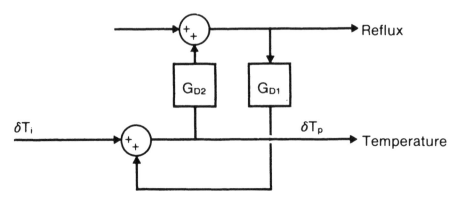

Figure 11.3.   Evaluation of interaction in a decoupled system.

decouplers do not form an immediate closed loop. In this situation the closed loop gain would have been:

$$0.018 * \frac{0.87}{0.96} * 4 * \frac{0.89}{0.87} = 0.067 \tag{11.48}$$

which would not have given any problems.

In case of a 2 × 2 system with feed forward (e.g., feed), the situation becomes more complex. In this case, signals would be picked up and added as shown in Figure 11.4.

A change in tower feed will manipulate both reflux and temperature. The decoupler to the reflux will change the reflux somewhat more because feed forward also causes the temperature to increase, which, if uncompensated, would cause the overhead concentration to change. The Smith Predictor signal is picked up at the controller outlet again, since feed forward and decoupler should control the process in such a way that feedback control is not necessary, therefore the predictor output should not change.

The decoupler signal from the reflux is added after the decoupler signal to the reflux in order to assure that a reflux change only causes a temperature change; any reflux change should not cause an additional change via the two decouplers.

DECOUPLING IN DYNAMIC RECONCILIATION

The 2 × 2 system in this case is described by the following static process equations:

$$x_D = x_{DO} + K_{11}R + K_{12}T \tag{11.49}$$

$$x_B = x_{BO} + K_{21}R + K_{22}T \tag{11.50}$$

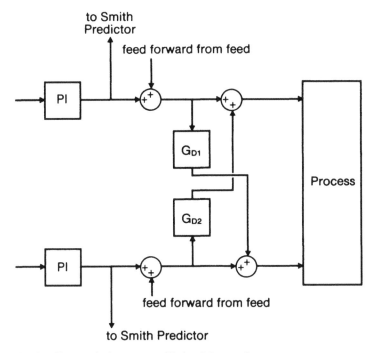

Figure 11.4. Decoupled system with feed forward.

The bias estimation is done using the dynamic equations:

$$x_{DO} = x_D - \hat{K}_{11}\hat{G}_{11}R - \hat{K}_{12}\hat{G}_{12}T \qquad (11.51)$$

$$x_{BO} = x_B - \hat{K}_{21}\hat{G}_{21}R - \hat{K}_{22}\hat{G}_{22}T \qquad (11.52)$$

The control equations are derived from equations (11.49) and (11.50) as:

$$R_{SP} = \frac{x_{D,SP} - x_{DO}}{\hat{K}_{11}} - \frac{\hat{K}_{12}}{\hat{K}_{11}} T_{SP} \qquad (11.53)$$

$$T_{SP} = \underbrace{\frac{x_{B,SP} - x_{BO}}{\hat{K}_{22}}}_{\text{feedback}} - \underbrace{\frac{\hat{K}_{21}}{\hat{K}_{22}} R_{SP}}_{\text{decoupler}} \qquad (11.54)$$

As shown in the case of feed-forward control, equations (11.53) and (11.54) only provide static decoupling. In order to achieve dynamic decoupling both equations have to be modified to:

$$R_{SP} = \frac{x_{D,SP} - x_{DO}}{\hat{K}_{11}} - \frac{\hat{K}_{12}\hat{G}_{12}}{\hat{K}_{11}\hat{G}_{11}} T_{SP} \qquad (11.55)$$

$$T_{SP} = \frac{x_{B,SP} - x_{BO}}{\hat{K}_{22}} - \frac{\hat{K}_{21}\hat{G}_{21}}{\hat{K}_{22}\hat{G}_{22}} R_{SP} \qquad (11.56)$$

The basic block diagram is shown in Figure 11.5.

Keep in mind that usually the reflux will be controlled by a secondary loop (flow control) and the temperature by another secondary loop (e.g., manipulating the steam flow). Consider the impact of a temperature change on x_D.

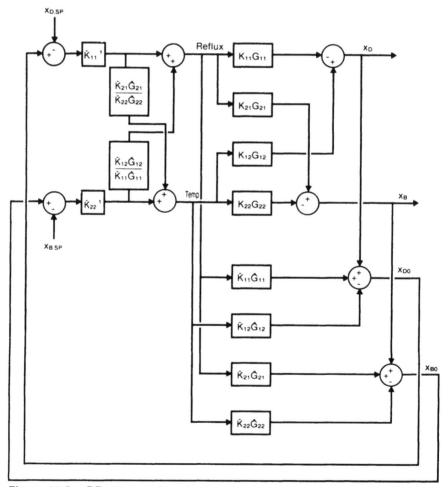

Figure 11.5. DR structure for a 2 × 2 system.

The overhead composition will be directly influenced by $K_{12}G_{12}$ and indirectly by $\hat{K}_{12}(\hat{G}_{12}/\hat{G}_{11}) \cdot (1/\hat{K}_{11}) \cdot K_{11}G_{11}$, therefore:

$$\delta x_D = K_{12}G_{12} - \hat{K}_{12}\frac{\hat{G}_{12}}{\hat{G}_{11}} \cdot \frac{1}{\hat{K}_{11}} \cdot K_{11}G_{11} \tag{11.57}$$

which will be zero in case of perfect models. If they are not perfect, decoupling control has to be augmented by feedback control. Define:

$$\left.\begin{array}{l} K_{11} = 0.03 \\ K_{12} = 0.1 \\ K_{21} = 0.06 \\ K_{22} = 1.0 \end{array}\right\} \tag{11.58}$$

$$G_{11} = \frac{0.11\ z^{-18}}{1 - 0.89\ z^{-1}}$$

$$G_{12} = \frac{0.133z^{-20}}{1 - 0.87\ z^{-1}}$$

$$G_{21} = \frac{0.04\ z^{-16}}{1 - 0.96\ z^{-1}} \tag{11.59}$$

$$G_{22} = \frac{0.13\ z^{-8}}{1 - 0.87\ z^{-1}}$$

The process models that are used in the control strategy are given by equations (11.33) to (11.36). The decouplers become therefore:

$$\frac{\hat{K}_{21}\hat{G}_{21}}{\hat{K}_{22}\hat{G}_{22}} = \frac{0.06}{1.0}\ \frac{0.04\ z^{-16}}{1 - 0.96\ z^{-1}}\ \frac{1 - 0.87\ z^{-1}}{0.133\ z^{-8}}$$

$$= 0.0185\ z^{-8}\ \frac{1 - 0.87\ z^{-1}}{1 - 0.96\ z^{-1}} \tag{11.60}$$

and

$$\frac{\hat{K}_{12}\hat{G}_{12}}{\hat{K}_{11}\hat{G}_{11}} = 4\ z^{-2}\ \frac{1 - 0.89\ z^{-1}}{1 - 0.87\ z^{-1}} \tag{11.61}$$

Note that these decouplers are the same as for the Smith Predictor as defined by equations (11.40) and (11.41). In the DR structure a decoupler produces an absolute value that is added to the absolute value of the controller output. It is possible to use incremental decoupling. In that case an incremental controller output also has to be calculated, which requires the control structure to be modified accordingly. In higher order systems dynamic decoupling becomes quite complex. Often static decoupling is used; the gains of the decouplers can easily be calculated from the gains of the process transfer functions. If this method does not give the desired results, other control techniques for multivariable systems should be used.

MATRIX APPROACH TO MULTIVARIABLE SYSTEMS

An extension of the Smith Predictor method to multivariable systems is given by Alevisakis and Seborg.[40] Since this extension gives a thorough review of all the mathematics involved, it will not be repeated here. Since the extension of the dynamic reconciliator to the multivariable case is not found in the literature it will be shown here. Equations (11.29) to (11.32) are written as:

$$
\begin{pmatrix} \delta x_D \\ \\ \delta x_B \end{pmatrix} = \begin{bmatrix} \dfrac{-0.03\ e^{-34S}}{1\ +\ 17S} & \dfrac{0.1\ e^{-38S}}{1\ +\ 14S} \\ \dfrac{0.06\ e^{-30S}}{1\ +\ 49S} & \dfrac{-1.0\ e^{-14S}}{1\ +\ 14S} \end{bmatrix} \begin{pmatrix} \delta R \\ \\ \delta T \end{pmatrix}
\tag{11.62}
$$

which can be rewritten in vector-matrix notation as:

$$
y = KG\ u
\tag{11.63}
$$

Equation (11.63) can also be written as:

$$
y = y_0 + KG\ u
\tag{11.64}
$$

in which y and u now represent absolute process values and y_0 is a bias, depending on the steady state and so forth.

Equation (11.64) can be written for a static situation as:

$$
y = y_0 + Ku
\tag{11.65}
$$

from which the setpoint for the control u can be derived:

$$
u_{SP} = [K]^{-1}\ (y_{SP} - y_0)
\tag{11.66}
$$

The bias can be estimated from equation (11.64):

$$
y_0 = y - KG\ u
\tag{11.67}
$$

Equations (11.66) and (11.67) describe the control system completely. The gain matrix K follows from equation (11.62):

$$K = \begin{bmatrix} K_{11} & K_{12} \\ K_{21} & K_{22} \end{bmatrix} = \begin{bmatrix} -0.03 & 0.1 \\ 0.06 & -1.0 \end{bmatrix} \tag{11.68}$$

Equation (11.66) can be rewritten, using equation (11.60), as:

$$\begin{pmatrix} R_{SP} \\ T_{SP} \end{pmatrix} = \frac{1}{K_{11}K_{22} - K_{12}K_{21}} \begin{bmatrix} K_{22} & -K_{12} \\ -K_{21} & K_{11} \end{bmatrix} \begin{pmatrix} x_{DSP} - x_{DO} \\ x_{BSP} - x_{BO} \end{pmatrix} \tag{11.69}$$

in which x_{DSP} and x_{BSP} are the setpoints for the top and bottom composition and x_{DO} and x_{BO} the bias for top and bottom according to equation (11.67). It should be noted that equation (11.69) gives static decoupling only and that control might be improved by adding dynamics to the decoupler.

PRACTICAL CONSIDERATIONS

In practice, especially in a 2 × 2 system, the relative gain will often not be evaluated, but interaction will be judged based on plant test runs. If, for example, a step in the reflux changes the overhead composition from 10% to 15% and the bottom concentration from 3% to 3.1%, a decoupler from reflux to bottom composition will not be implemented. The relative gain matrix for a 2 × 2 system does not provide information on whether one or two decouplers are required.

This can, however, be evaluated on the basis of equation (11.24). It can be seen from this equation that a reflux change affects the overhead concentration via a coefficient 0.03 and the bottom composition via a coefficient 0.06. Since these numbers are of the same order of magnitude we do need a decoupler from reflux to bottom composition. However, the temperature affects the bottom composition via a coefficient 1.0 and the top composition only via a coefficient 0.1. Since the latter coefficient is ten times smaller than the first one, a decoupler from temperature to overhead composition is not required. As said before, 3 × 3 and higher order systems need to be evaluated carefully. Use static decouplers if possible and examine carefully if partial decoupling could solve your control problem. As a rule of thumb, do not add more than one decoupler signal to the output of a particular controller.

EXAMPLES

The example that is simulated on the diskette is the process as described by equations (11.29) to (11.32).

File SP22 is simulating control of the process using Smith Predictors and decouplers. File DR22 is simulating the controlled process using dynamic reconciliators and decouplers. The block diagram for the case in which Smith Predictors are used was shown in Figure 11.2.

Process Description

The process equations are written in terms of deviation variables:

$$K_{11}G_{11} : R_{df,k} = -0.00333 \ (R_{k-18} - 280) + 0.89 \ R_{df,k-1} \qquad (11.70)$$

$$K_{12}G_{12} : T_{df,k} = 0.0133 \ (T_{k-20} - 210) + 0.87 \ T_{df,k-1} \qquad (11.71)$$

in which $R_{df,k}$ = delayed and filtered reflux (output) at time k
$T_{df,k}$ = delayed and filtered temperature (output) at time k
R = reflux, process input
T = temperature, process input

The top concentration x_D now becomes:

$$x_{D,k} = T_{df,k} - R_{df,k} + 10 \qquad (11.72)$$

The steady state values of reflux, temperature, and concentration are 210, 280, and 10 respectively.
The other process equations are:

$$K_{21}G_{21} : R_{df,k} = 0.0024 \ (R_{k-16} - 280) + 0.96 \ R_{df,k-1} \qquad (11.73)$$

$$K_{22}G_{22} : T_{df,k} = -0.133 \ (T_{k-8} - 210) + 0.87 \ T_{df,k-1} \qquad (11.74)$$

with the equation for the bottom concentration:

$$x_B = T_{df,k} + R_{df,k} + 20 \qquad (11.75)$$

Smith Predictor Equations

The equation for the overhead loop is:

$$y_k = 0.0033 \ (x_{k-1} - x_{k-18}) + 0.89 \ y_{k-1} \qquad (11.76)$$

with y_k = output of the predictor at time k
x_k = input to the predictor at time k

Similarly, the equation for the bottom loop is:

$$y_k = 0.133 \ (x_{k-1} - x_{k-8}) + 0.87 \ y_{k-1} \qquad (11.77)$$

Controller Equations

The error e on which control action is based is:

$$\text{error} = -\text{setpoint} + \text{analyzer reading} - \text{Smith Predictor output}$$

which translates for the overhead loop into:

$$e_k = -sp_k + x_{D,k} - y_k \tag{11.78}$$

and for the bottom loop into:

$$e_k = -sp_k + x_{B,k} - y_k \tag{11.79}$$

The output u_k of the PI controller is:

$$u_k = u_{k-1} + K_c \left[e_k - e_{k-1} + \frac{\Delta t}{\tau_i} e_k \right] \tag{11.80}$$

with K_c = controller gain
τ_i = integral time, min
Δt = algorithm execution interval, min

The input to the Smith Predictor is calculated from:

$$x_k = x_{k-1} + \underbrace{(u_k - u_{k-1})}_{\text{change in output PI controller}} \tag{11.81}$$

The reflux can then be calculated from:

$$R_k = R_{k-1} + \underbrace{(u_{d,k} - u_{d,k-1})}_{\text{change in output decoupler}} + (u_k - u_{k-1}) \tag{11.82}$$

and the temperature from:

$$T_k = T_{k-1} + \underbrace{(u^*_{d,k} - u^*_{d,k-1})}_{\text{change in output decoupler}} + (u_k - u_{k-1}) \tag{11.83}$$

The decoupler outputs are calculated using equations (11.40) and (11.41).
With the decouplers active a change in the setpoint of one control loop does

not affect the composition of the other loop. The interaction in the system is shown in Figure 11.6. In this case the decouplers were inactive.

DYNAMIC RECONCILIATION

This control structure is very similar to that of a Smith Predictor. The term x_{k-1} in the Smith Predictor is deleted and the PI controller is replaced by the inverse of the process gain, as shown in Figure 11.5.

CHAPTER 12

Constraint Control

So far, control in which there is no active constraint was discussed. It is not uncommon, however, to encounter situations where constraints become violated during process operations. Design of a control scheme should take these situations into account, and the result is often a more complex control scheme than if there were no constraints. If constraints are not taken into account, they may be violated, which may result in serious loss or damage to equipment and the environment. Two types of constraints can be distinguished:

- hard constraints

- soft constraints

Each of these types of constraints will be discussed in more detail. Once the supervisory control scheme is designed and all constraints are properly dealt with, there may still be a number of degrees of freedom that are not used for control. These remaining process variables can then be used for optimization. This is discussed in the next chapter.

HARD CONSTRAINTS

Hard constraints are constraints that should not be violated because of operational reasons or cannot be violated because of physical reasons. A typical example of a hard constraint is a valve position which cannot exceed 0 and 100%. This is a physical constraint that, upon violation, can result in loss of control. If the controller would continue to output once the valve has reached a limit, the controller would saturate or wind up. Most modern electronic controllers have antiwind-up circuits that inhibit integral action once the valve reaches the open or closed position. In a cascade system the primary controller can also be protected against wind-up by monitoring the deviation between secondary setpoint and process variable. An effective way of preventing wind-up is to maintain a constant secondary setpoint as long as the deviation between secondary setpoint (SP) and measured value (PV) is larger than a predefined limit. Only changes are allowed that move the setpoint away from the constraint.

Most supervisory control systems offer standard control algorithms that inhibit integral action in the primary controller as long as the secondary controller is in a wind-up state. (See Figure 12.1). It is of ultimate importance to use this feature if available. If it is not available, it is better to write your own control algorithm and build in the antiwind-up protection.

In the case where feed-forward control action is added to the output of the primary controller the situation is somewhat more complex. Also in this case, however, the deviation between secondary setpoint and measurement could be used to arrest output from the primary to the secondary controller.

Another example of hard constraints are equipment constraints. Some protection can be obtained by limiting the controller output between high and low limits although changes in process operation may change the relationship between valve position and equipment constraint. It will therefore be better to measure the constrained variables and use them in a control scheme. An example is a distillation tower where the reboiler is on temperature control. The tower vapor flow should remain within a low and high limit in order to avoid weeping or flooding. Both phenomena are rather irreversible, e.g., when a tower starts flooding, operating conditions have to be changed considerably in order to return to a normal operating mode. A control scheme that protects against high and low vapor flow is shown in Figure 12.2. If the tower starts flooding, the differential pressure controller with the high setpoint will reduce the steam flow and take over control from the temperature controller.

Another example of a hard constraint is the surge point of a compressor. It is not acceptable to exceed the surge limit, and, as such, the compressor has to be protected against the flow becoming too low.

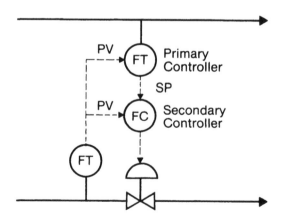

Figure 12.1. Feedback of secondary measurement to prevent wind-up.

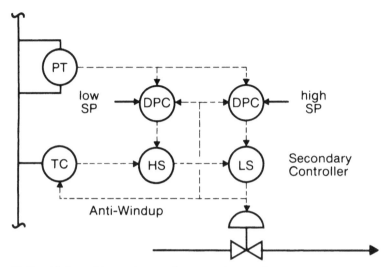

Figure 12.2.   High and low selectors in protecting equipment.

SOFT CONSTRAINTS

Soft constraints are limits that can usually be violated for short periods of time without causing serious problems. An example is the product of a distillation tower which is blended with other components and is then sent to a hold-up tank. Minor variations in tower product quality can be blended out and will be further smoothed out in the hold-up drum. Another example is the tube temperature in furnaces. A higher than usual tube temperature is often allowed provided the period over which it occurs is short. Soft constraints are approachable from both sides of the operating region in contrast with the compressor surge constraint, which is always approached from one side of the constraint. In general, a constraint can be characterized by an expression of the form:

$$\int_0^T \Phi(x) \, dt \leq c_{max} \tag{12.1}$$

in which Φ is a linear or nonlinear function of one or more process variables x; T is an averaging period or the period of optimization and c_{max} the limiting value. Some constraints classified according to equation (12.1) are shown in Figure 12.3.[41] In the graph the degree of nonlinearity Φ is plotted vs the averaging period. In this graph various phenomena occurring in chemical plants can be plotted. It should be noted that when the value of T is equal to zero, the horizontal scale no longer has significance.

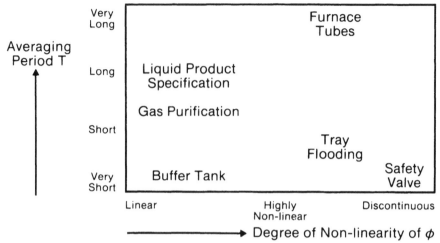

Figure 12.3. Classification of constraints. (Reproduced with permission from the American Institute of Chemical Engineers.)

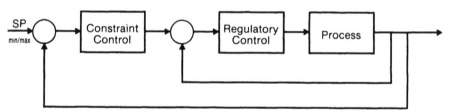

Figure 12.4. Single variable constraint control.

CONTROL NEAR ONE CONSTRAINT

Process operation is often optimal near a constraint. In that case there is the desire to maintain the process near the constraint despite process upsets. Figure 12.4 shows the diagram for control of a single variable near a constraint.

This diagram looks very much like a cascade control system. The constraint controller can be an integral controller only, which slowly drives the regulatory setpoint. The changes in the regulatory setpoint are so slow that the regulatory control loop has sufficient time to adjust to the changes in its setpoint. The integral time should be set as:

$$\tau_i = K_p \Delta t \tag{12.2}$$

in which K_p = process gain
         Δt = tag execution interval

The tag execution interval should be least equal to the dead time plus three times the time constraint:

$$\Delta t \geq \theta + 3\tau \qquad (12.3)$$

in which θ = process dead time
τ = process time constant

It is obvious that this type of constraint control will give a very slow response although it is applied, for example, in tower pressure minimization as shown in Figure 12.5. A valve position controller VPC adjusts the pressure setpoint. The intention is to hold the condenser control valve in its fully open position (on the maximum constraint) in order to minimize the pressure and consequently minimize energy consumption. The pressure controller PC has proportional and integral action in order to eliminate process upsets quickly. The integral valve position controller moves the pressure setpoint to its optimal long-term value. Wind-up of the VPC is prevented by using the pressure measurement as reset feedback.

Constraint control often creates a very nonlinear objective function. To make this clear, consider control of the acetylene concentration at the outlet of an acetylene converter. If the target acetylene concentration is 25 ppm and the actual concentration drops to 15 ppm, there is obviously the need to bring it back to setpoint. However, if the concentration increases to 35 ppm, the situation is more serious and the concentration has to be brought back to target much faster in order to avoid off-spec material.

This creates a situation where even control action may become nonsymmetric: upon a negative offset the controller acts faster than on a positive offset. In the literature there are at least ten different nonlinear controllers proposed to improve control near a constraint. From practical experience it was found

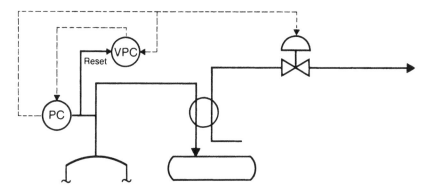

Figure 12.5.   Tower pressure minimization.

that all of these controllers offer only limited success. One controller that was found to be useful for the situation just outlined was one in which the gain was tripled for positive offsets. Another interesting approach is given in Westerlund et al.[42] In this case the authors do not try to develop a nonlinear controller but update the controller setpoint based on past process values. A process is operated against a soft constraint, i.e., only during a certain period is the controlled variable y allowed to be outside the desired quality limit y_M. The setpoint of y at time k is denoted $y_{SP,k}$. A natural choice for the setpoint is given by:

$$y_{SP,k} = y_M \pm \Delta y_{M,k} \tag{12.4}$$

where $\Delta y_{M,k}$ is a "confidence" interval. The - or + sign corresponds to an upper or lower quality limit. The confidence interval can be given by:

$$\Delta y_{M,k} = Fs_{H,k} \tag{12.5}$$

where F is a constant factor, or more precisely, it is obtained from the density function of the stochastic process H(k), and s_H is the estimated standard deviation of H(k):

$$H(k) = y_k - y_{SP,k-1} \tag{12.6}$$

If slow variations in the estimation of H are allowed, the variance of H can be approximately computed by giving old values of $H^2(k)$ an exponentially decreasing weight:

$$s^2_{H,k} = \frac{\sum_{i=0}^{k} \lambda^{k-i}[y_k - y_{SP,k-1}]^2}{\sum_{i=0}^{k} \lambda^{k-i}} \tag{12.7}$$

where λ is a weighting factor, $0 < \lambda < 1$. Equation (12.7) can be approximated by:

$$s^2_{H,k+1} = \lambda s^2_{H,k} + (1 - \lambda)[y_{k+1} - y_{SP,k}]^2 \tag{12.8}$$

Equations (12.4), (12.5), and (12.8) give a simple method for on-line updating of the setpoint. The value of F can be obtained from the assumed normal distribution of y and is given in Table 12.1 (extracted from tables for cumulative normal distribution):

Table 12.1. Value of F vs Percent of Process Values Allowed above the Constraint

Percent of values allowed above constraint	F
1.0	2.33
1.5	2.17
2.0	2.06
2.5	1.96
3.0	1.88
4.0	1.75
5.0	1.65
6.0	1.56
7.0	1.48
8.0	1.41
9.0	1.34
10.0	1.28

Example

F	$= 1.96$	(2.5% of value above y_M)
y_M	$= 10.5$	(limit)
y_{sp}	$= 10.0$	(starting point)
s^2_H	$= [(y_M - Y_{SP})/F]^2$	(from eqn (12.4) and (12.5))
λ	$= 0.98$	(exponential weight)
y_{SP}	$= y_M - Fs_H$	
y	$=$ new measurement	
s^2_H	$= \lambda * s^2_H + (1 - \lambda) * (y - y_{SP})^2$	

which will adjust the setpoint as time progresses. The simulation using this example is given in file CONSTR. A first order process with a small delay was simulated. Figure 12.6 shows control of this process. The setpoint is slowly increased in order to operate close to the constraint. Because noise is added to the process signal, a certain distance from the constraint is maintained.

CONTROL NEAR MULTIPLE CONSTRAINTS

The previous section presented two techniques of driving a process variable against a constraint. However, sometimes more than one process constraint can limit process operation and more advanced control is required. This section will deal with that situation. The process variables of interest should be monitored or calculated. A detailed plant model can be used where necessary to estimate constraints which cannot be measured directly. The control system must determine which of the process variables is the active constraint and limits plant operation. Plant production can then be maximized by adjusting

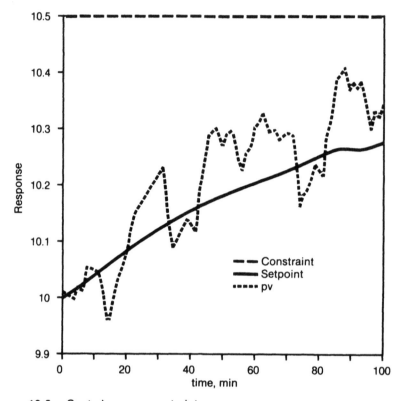

Figure 12.6.   Control near a constraint.

the predetermined manipulated variable in order to drive the plant against the
active constraint. A detailed process analysis is usually required to determine
which process variables have to be considered. Three approaches to the constraint control problem will be discussed:

• steady state constraint control

• dynamic constraint control using a single PI controller

• dynamic constraint control using multiple PI controllers

STEADY STATE APPROACH

If it is not required to control the process tightly near the constraints, a steady state approach will suffice. In this case the manipulated variable is increased by a fixed amount Δu, after which the control system allows the process sufficient time to settle out. If no constraints are violated, the manipulated variable is again increased by the same amount. This procedure is repeated until the process variable reaches a constraint. If a constraint is violated the control system decreases the manipulated variable by a fixed amount, which should be about twice as much as the change made when moving toward the constraint.

It is obvious that this type of constraint control is not very tight because it ignores dynamics completely; however, a large portion of the credits of constraint control can be captured this way with relatively minor effort.

Dynamic Constraint Control Using Single PI Controller

This situation is very similar to the one discussed in the previous section. It can be used here when the dynamics of the different constraints are similar. Figure 12.7 shows the approach. Each constraint has a dedicated integral-only controller. As in Figure 12.4, the PV of the controller is the constraint variable. Since the constraints can be imposed upon different process variables they are normalized according to:

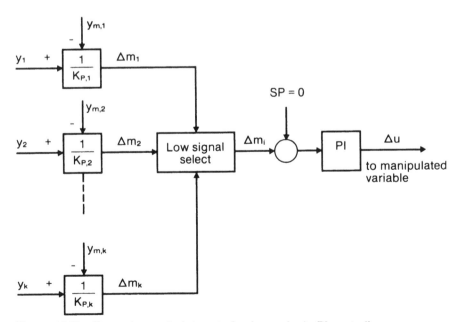

Figure 12.7. Dynamic constraint control using a single PI controller.

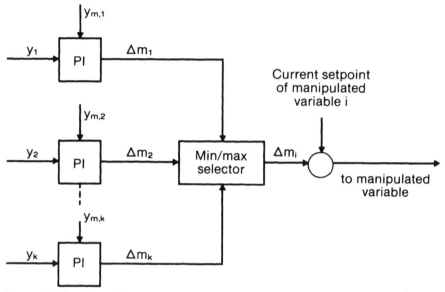

Figure 12.8.   Dynamic constraint control using multiple PI controllers.

$$\Delta m_i = \frac{y_i - y_{m,i}}{K_{P,i}} \qquad (12.9)$$

in which Δm_i = change in manipulated variable, required to bring the ith constraint variable to its limit in one time step
y_i = current value of the ith constraint variable
$y_{m,i}$ = constraint value of the ith constraint variable
$K_{P,i}$ = process gain = $\delta y_i/\delta u$
u = manipulated variable

All values of Δm are compared and the minimum is selected because any
larger value of Δm would violate one or more of the constraints. The integral
action of each integral-only controller is calculated using equation (12.2). The
selected output Δm_i is the PV of the PI controller. The setpoint of this controller is zero. The purpose of this is to output the selected Δm_i slowly to the
process, so that when a constraint is be violated, the process will not be upset.
The single PI constraint controller is simple to implement and proven to be
quite adequate in situations where the dynamics of the different constraints are
similar.

Dynamic Constraint Control Using Multiple PI Controllers

When the constraints have different dynamics, this approach is required for good constraint control. The scheme is shown in Figure 12.8. In this case each constraint has a separate PI controller calculating Δm_i. A selector picks the largest Δm_i if a constraint is violated and has to be relieved, and the smallest Δm_i if the process can still be driven toward the constraint. The PI controllers have proportional action based on PV rather than error. The output Δm_i is added to the current setpoint of the manipulated variable. A drawback of this approach is that it is rather sensitive to noise. If one or more of the signals y_1 to y_k is noisy, the best thing to do is to eliminate the source of the noise or filter it out.

Complex Situations

There may be situations where multiple constraints are defined and multiple variables can be manipulated to drive the plant toward the constraints. In that case none of the approaches that were discussed are suitable. However, linear or nonlinear programming can be used to optimize plant operation by operating close to the constraints.

CHAPTER 13

The Design, Adjustment, and Application of Process Models

Sometimes one gets the impression that with the increase of computer memory, models also increase in complexity. Obviously the type of model and the complexity are very much related to the application. In case one wants to predict the temperature profile in a chemical reactor, it is not sufficient to look for a simple relationship between reactor outlet temperature and the input variables such as flow and temperature—the model has to be more detailed. For on-line use, however, the model must be simple. This is important not because computing memory and time have to be minimized, but because it is important for quick error detection and adaptability of the model to changes in process equipment and increased process knowledge. Hence, there will be continuous interaction between the application and the design resulting in an improved model that better approaches reality. An on-line model should also have a minimum number of unknown parameters. This is essential for the adjustment to changes in process behavior, for example, decay in catalyst activity or decrease in heat transfer due to fouling. Both these undesired changes and increased process knowledge have to be implemented in the process model.

TYPES OF PROCESS MODELS

The most simple "process models" are not more than a recipe of desired values. Via a test run the best settings have been determined, and those settings are now prescribed for future operation, for example, in the form of tolerance limits for the input and output variables of the process:

$$u_{min} \leq u \leq u_{max} \tag{13.1a}$$

$$y_{c,min} \leq y_c \leq y_{c,max} \tag{13.1b}$$

157

In this equation u are the adjustable input variables of the process; y_c are the controlled output variables. The checking and supervision can be done by an on-line computer. If the process operation has a dynamic character, the tolerance limits will change as a function of time. An example is the cooling of a steel plate in which a certain temperature profile has to be maintained. In this case the on-line computer can be very beneficial. In a model in a real sense, output variables of the process are functions of the input variables. It is important to distinguish between static and dynamic process models. Static models take the following form (Figure 13.1):

$$y = f_y(u^*, \bar{w}, \alpha)$$

(13.2a)

$$q = f_q(y, u^*, \bar{w}, \alpha)$$

(13.2b)

in which u^* are the known (measurable) input variables; $\bar{w}$ the nonadjustable (nonmeasurable) input variables (disturbances); α the process parameters; y the automatically (on-line) measured output variables; and q the nonautomatically (off-line) measured output variables. Since w cannot be measured, the average value is substituted in equation (13.2).

In Figure 13.1 a distinction has been made between y, the automatically measured (on-line) variables, and q the nonautomatically measured (off-line) output variables, together representing the output variables y^*. The off-line variables are, in general, the product qualities which, after sampling (S), are analyzed in the lab with a delay (D). A possible application of on-line models is the estimation of q on the basis of on-line measurements of u^* and y. In Figure 13.1 a distinction is made between nonadjustable known input variables u_v and adjustable known variables u, and between controlled and noncontrolled output variables, y_c and y_m respectively.

Dynamic models can have all kinds of forms. An important category of models is based on the state vector x. They can be represented by a set of first order differential equations:

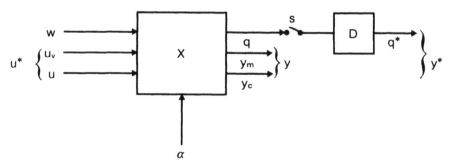

Figure 13.1.   General block diagram of a process.

$$\frac{dx}{dt} = f_c(x,u^*,\alpha,\bar{w}) \qquad (13.3a)$$

$$y = g_y(x,u^*,\alpha,\bar{w}) \qquad (13.3b)$$

$$q = g_q(x,u^*,\alpha,\bar{w}) \qquad (13.3c)$$

In many cases y can be represented as a subgroup of x. In an on-line process computer the discrete versions of equation (13.3) are used:

$$x(k + 1) = f_d[x(k),u^*(k),\alpha,\bar{w}] \qquad (13.4a)$$

$$y(k + 1) = g_y[x(k+1),u^*(k+1),\alpha,\bar{w}] \qquad (13.4b)$$

$$q(k + 1) = g_q[x(k+1),u^*(k+1),\alpha,\bar{w}] \qquad (13.4c)$$

THE DESIGN OF A PROCESS MODEL

The design of a theoretical model usually starts with writing down the conservation laws resulting in momentum, material, and energy balances.

When the process is divided into a number of ideally mixed sections, the result for dynamic models will be first order differential equations such as equation (13.3a). There are also relationships between physical parameters (e.g., the gas law), kinetic equations (e.g., Arrhenius equation for chemical reactions), and empirical relationships (e.g., pressure drop due to flow in a pipeline).

Often a "complete" dynamic model is too complicated to handle. One then uses quasi-theoretical models, in which, for example, mixtures of many components are approached by a binary mixture.[43] For chemical reactors the (often unknown) reaction mechanism is then simplified to a first order mechanism.

Static Empirical Models

Theoretical models have to be compared to empirical data. It is possible that only the values of parameters have to be adjusted; in other words, the theoretical model can be used without modification. In other cases an adjustment of the model may be necessary.

It is also possible to postulate just a general form of the model, which only is validated after implementing the measurements. Usually polynomials are used:

$$y_k = \alpha_0 + \alpha_1 u^*_1 + \ldots + \alpha_{11}(u^*_1)^2 + \alpha_{12}u^*_1u^*_2 \qquad (13.5)$$

The coefficients α_0, α_1, α_{11}, . . . can be determined by using regression analysis. One should be aware that the number of terms may not become too large, thereby adjusting the model to only a particular set of measurement results.

Very attractive are computer programs for regression analysis which actually build the equation by adding additional terms. The significance of the improvement resulting from the addition of a new term is weighed and as soon as the improvement is insignificant, the process is stopped. Harder et al.[44] found that usually five to ten terms provide the required results. Experimental results can be obtained in different ways: from normal process operation, special test runs, and measurements on a laboratory or semitechnical scale. A special case of significant importance is the approximation of a complicated off-line model by a simple empirical model. The parameters of the latter model are determined from runs with the off-line model. Normal process operation often has the disadvantage that it has been too steady. This can be avoided by making special test runs, but this is usually fairly expensive.

Dynamic Empirical Models

Polynomials can be extended with values from the past, hence the dynamic behavior is taken into account (compare with equation 13.5):

$$y_k = \alpha_0 + \alpha_1(k)u^*_1(k) + \alpha_1(k-1)u^*_1(k-1) +$$
$$\dots + \alpha_2(k)u^*_2(k) + \dots + \alpha_{11}(k)[u^*_1(k)]^2 + \qquad (13.6)$$
$$\alpha_{11}(k-1)[u^*_1(k-1)]^2 + \dots$$

The number of parameters, however, increases very rapidly; hence, this method has to be applied with ultimate care. If the dynamic behavior is known to a certain extent, it is possible to reduce the number of terms by the right selection of the time interval.

Two techniques which received considerable alteration in the literature are dynamic matrix control (DMC) and identification and command (IDCOM). DMC[45-49] is a multistep-ahead predictive control strategy. The strategy is formulated by using a discrete convolution representation for the process model (i.e., step weights). The algorithm uses measurements of the current error and past control inputs. The discrete convolution model for any single input–single output combination is:

$$y(k + 1) = \sum_{i=1}^{\infty} a_i \Delta u(k - i - 1) + y_0 + d(k + 1) \qquad (13.7)$$

in which k = time
y_0 = initial output condition
$\Delta u(k)$ = change in input
$y(k)$ = absolute value of output at time k

$d(k)$ = plant/model mismatch at time k
a_i = unit step response coefficients

IDCOM is a model-predictive heuristic control technique.[50,51] The plant is represented in this case by its impulse responses which will be used by a control computer for long-range prediction.

Both techniques find industrial application in control of multivariable processes.

Practical Examples

In practice one will strive to combine theoretical and empirical data. If certain theoretical equations are reliable, replacement by general polynomials will usually result in an undesired increase in the number of parameters. Below some examples are given from the literature.

Bray et al.[52] studied the conversion of carbon monoxide and steam into carbon dioxide and hydrogen in a chemical reactor (water gas reaction):

$$CO + H_2O \rightarrow CO_2 + H_2 \qquad (13.8)$$

Based on a kinetical analysis of the reaction a complicated algebraic equation was developed, which was approximated by series development. The series could be simplified by taking the limited operability range of the process variables into account. Finally the following equation was obtained:

$$y = \alpha_0 + \alpha_1 u_1 + \alpha_2 u_1 u_2(k-1) + \alpha_3(u_1)^2$$
$$+ \alpha_4(u_1)^2 u_2 + \alpha_5(u_1)^2(u_2)^2 \qquad (13.9)$$

in which y is the carbon monoxide concentration in the effluent, u_1 the CO/H_2O ratio in the feed, and u_2 the feedrate. The number of terms is much lower than that of a general fourth order polynomial. The dynamics were taken into account by adding two weighed averages from past values of u_1 and u_2. The first is more or less representative for the dependence of y of the average temperature in the reactor (time delay approximately 0.5 hours) and the second represents the dependence of the temperature in the reactor of the conversion (time delay approximately 3 hours). The total number of terms now became $\alpha_0 + 3.5 (\alpha_1 \ldots \alpha_5) = 16$, which is still fairly high.

Roberts and Laspe[53] and Duncanson and Prince[54] have developed on-line models for cracking furnaces. They introduce a variable called the cracking severity, which is a function of the input variables. This parameter determines the concentration of the components in the effluent (Figure 13.2). The introduction of this parameter significantly reduced the number of model equations.

Duyfjes and Van der Grinten[43] approximated the static behavior of multi-

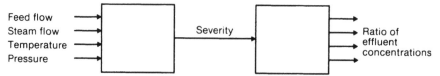

Figure 13.2.   Model structure for a cracking furnace.

component distillation columns by the Shinskey[55] version of the Fenske-Underwood-Gilliland model, which gives an approximation of the behavior of binary distillation columns. Also an intermediate variable is introduced: the degree of separation S:

$$S = \frac{x_D(1 - x_B)}{x_B(1 - x_D)}$$ (13.10)

in which x_D is the concentration of the most volatile main component in the distillate, x_B the same for the bottom product, and S a function of the relative vapor load:

$$S = a + b\frac{V}{F}$$ (13.11)

in which V is vapor flow and F is the feed flow.

This model has only two parameters, a and b; the approximation is very good, however, for a number of columns.

Bertoni et al.[56] developed an empirical model for an electrochemical process in which the voltage is calculated from the current I, the temperature T, and the concentration c. The model has the following form:

$$y = (\alpha_0 + \alpha_1 I) + (\alpha_2 + \alpha_3 I)T + \alpha_4 T^2 + \alpha_5 T^3$$
$$+ \alpha_6 T^4 + \alpha_7 c$$ (13.12)

in which y represents a change in voltage. The parameters $\alpha_0 \ldots \alpha_7$ are selected in such a way that the model fits the experimental results in an area around the operating point.

Kerlin et al.[57] have developed dynamic empirical models for power plants. The models have the form of equation (13.6), in which the parameters are identified by means of a suitable identification method.

For a small scale fermentation process Halme and Holmberg[58] developed a simple dynamic model based on the Monod equation and equations for biomass and substrate. After parameter adjustment the model was used for digital control and optimization of a commercial fermentation process.

Based on physical considerations and experimental results, Hammer,[59] developed a simple model for the mixing of four powdery components. The transfer function from storage tank to measuring device for each component:

$$y_k = \frac{bz^{-f-1}}{1 - az^{-1}} x_k \qquad (13.13)$$

in which y_k is the output at time k; x_{k-f-1} the input at time k–f; f is the delay time; and z the backward shift operator given by $y_{k-1} = z^{-1}y_k$. The mixing process is controlled by a digital computer in which PI controllers are programmed. The computer adjusts controller parameters in such a way that the square of the deviation between measured and desired mixing ratio is minimal over time.

ADJUSTMENT OF ON-LINE MODELS

On-line models usually have to be adjusted for gradual changes in the process. In fact the distinction between process variables and parameters is not absolute: parameters are process variables that change slowly with respect to the dynamic behavior of the process variables. Adjustment generally means that in simple cases Figure 13.1 has to be reversed[60] (see Figure 13.3). When the models are more complicated, it will be almost impossible to reverse the equations. In that case one can use model adjustment techniques based on the method shown in Figure 13.4. Based on the measured input variables the output variables are calculated by using the model. Possible deviations between model values (Y_{model}) and process values ($Y_{process}$) are decreased by an iterative algorithm. The norm is a suitable measure for the size of the deviations, for example:

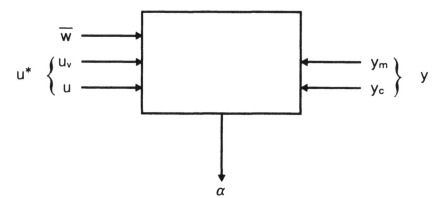

Figure 13.3. Updating of process parameters.

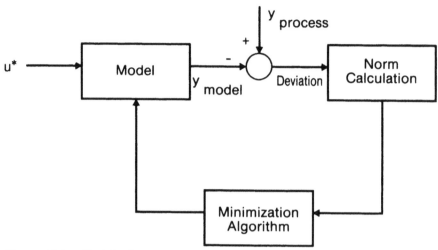

Figure 13.4.   Model adjustment method.

$$\Sigma^k \, (Y_{k,model} - Y_{k,process})^2 \qquad\qquad (13.14a)$$

or

$$\Sigma^k \, | \, Y_{k,model} - Y_{k,process} \, | \qquad\qquad (13.14b)$$

where index k has to be summed for all output variables. The selection of the norm is not arbitrary; a certain norm is optimal for a certain application.

In case of static models one has to make sure to have a large number of measurements obtained over a long period of time in order to average dynamic effects. The measurements also have to be consistent, e.g., material balances should be consistent. If necessary, they have to be made consistent via a least-square method.[61,62] The updating preferably has to be done by an on-line computer. If the model is linear in the parameter values, a special version of linear regression can be used.

With this method the continuous inversion of large matrices is avoided. Through the introduction of a weighing coefficient there is a decreasing influence of the past.

Automatic updating also has its disadvantages: if there is a change in process operation (for example, by switching to another feed), it is better to revert back to parameter values obtained in a previous run with the same feed. Large disturbances, repairs to pieces of equipment, and so forth can complicate the situation further.

For dynamic models a large variety of parameter estimation techniques is available. It would fall beyond the scope of this book to explain them in detail.

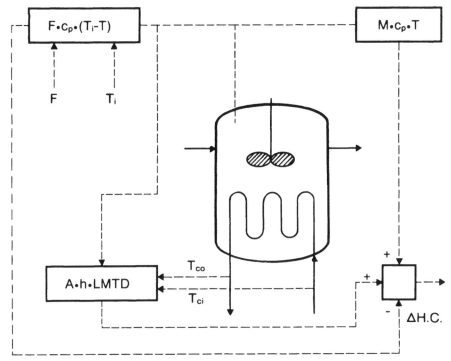

Figure 13.5. Estimation of conversion.

The proceedings of the IFAC symposia about system identification[63] and the work of Eykhoff[64] give detailed information about this subject.

APPLICATION

Process models can be suitably used in the estimation of process parameters which cannot be measured or are difficult to measure. The models are relatively simple and dynamic compensation is usually no problem.

An example is the estimation of the conversion in a reactor.[41] Figure 13.5 shows the principle that is based on the dynamic energy balance of the process:

$$Mc_p \frac{dT}{dt} Fc_p(T_i - T) - Ah(LMTD) + \Delta HC \qquad (13.15)$$

in which M = reactor mass contents, kg
c_p = specific heat (J/kg)
T = temperature (°K)
t = time (sec)
F = feed flow (kg/s)

T_i = temperature feed (°K)
A = heat transfer area cooling coil (m²)
h = heat transfer coefficient (W/m²sec)
LMTD = logarithmic mean temperature difference between reactor contents and cooling coil (°K)
ΔH = heat of reaction (J/kg)
C = conversion (kg/sec)

If the flow of cooling water can be measured, the term with the logarithmic mean temperature difference can be replaced by

$$F_c c_{pc}(T_{co} - T_{ci}) \tag{13.16}$$

in which F_c = cooling water flow (kg/sec)
c_{pc} = specific heat cooling medium (J/kg °K)
T_{co} = exit temperature (°K)
T_{ci} = inlet temperature (°K)

Another example is the estimation of product qualities. These can usually only be determined at large time intervals. A dynamically favorable value would benefit the operators. The model can then be adjusted based on a measurement of the product quality. This principle has been applied in a number of occasions in crude distillation processes.[65,66] In these processes a number of intermediate products are produced, which are characterized by initial and final boiling point. Process temperatures in the right places are a measure for the average of the final boiling point of an intermediate product and the initial boiling point of the next intermediate product. The difference of these latter boiling points can be estimated from process flows.

Crude oil distillation processes also give another example.[65] The internal flows within a distillation column have to be maintained within certain limits. They can be determined through measurement of externals flows, temperatures, and so forth.

STATE ESTIMATION IN DYNAMIC MODELS

In dynamic models state estimation is often important. Elements in the state vector can be representative for relevant process variables such as product quality. Figure 13.6 shows the structure of a state estimator (for symbols, see also Figure 13.4). Based on the estimated state x (k/k) at time k and on measured values of the input variables y(k), a prediction is made one step ahead in time using the model equation:

$$\hat{x}(k + 1/k) = f_d[\hat{x}(k/k), u^*(k), \bar{w}] \tag{13.17a}$$

$$\hat{y}(k + 1/k) = g_y[\hat{x}(k/k), u^*(k), \bar{w}] \tag{13.17b}$$

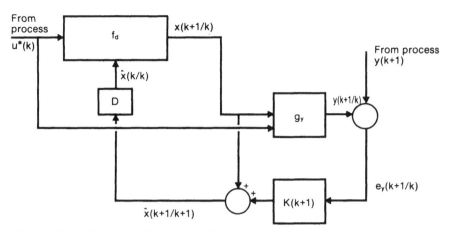

Figure 13.6.   Structure of a state estimator.

The predicted output vector at time $k + 1$ is compared to the measured output vector $y(k + 1)$. The difference vector, the so-called innovation $e_y(k + 1/k)$, is used for correction of the predicted state. For a well-tuned state estimator the innovation is small; a linear correction is sufficient in that case:

$$\hat{x}(k + 1/k + 1) = \hat{x}(k + 1/k) + K(k + 1)\, e_y(k + 1/k) \qquad (13.18)$$

in which $K(k + 1)$ is the correction matrix.

Notice the difference between $x(k + 1/k)$ and $x(k + 1/k + 1)$. The former is the prediction starting at time k, the latter is the estimation based on new information that is available at time $k + 1$. D is a delay with one time step: it is equal to the waiting time between calculations at different points in time. If w and measurement errors can be ignored, the state estimator is often called observer.[67,68] Theoretically the state can be observed ideally in a limited number of time steps. If w and/or the measurement errors cannot be ignored, the state estimator is usually called Kalman filter.[69,70] Seborg et al.[70] applied the filter in a computer controlled evaporator. The technique is used to estimate the state variables concentration and temperature, which are needed for optimal control. Powell et al.[71] and Church[70] apply Kalman filters in the processing of pulp. By using the filter the pulp composition is estimated, and the model is corrected on the basis of measurements. From the literature one gets the impression that the Kalman filter finds many applications on a laboratory or semitechnical scale; however, fewer applications are found in industry. This situation will most likely change in the near future.

APPLICATIONS OF PREDICTIONS

Process models can also be used to predict future values of process variables. This can be very important in batch operation or in processes with a large residence time. An example is the copolymerization of styrene and butadiene in series of mixed reactors (total residence time about ten hours). The ingredients are mixed in the first reactor, the reaction is initiated in the first or second reactor, and reaction proceeds in the other reactors. A model that will predict the final result, based on measurements in the first reactor, will be a valuable tool for the operator. Other examples of prediction models can be found in Powell et al.,[71] Church,[70] and Amrehn.[72]. The first two authors used the following model for the prediction of the quality of pulp:

$$y = y_0 + a(T - T_0) + b(C - C_0) + c(P_H - P_{HO})$$
$$+ d(x - x_0) + e(\tau - \tau_0)$$

(13.19)

in which y is the predicted variable, e.g., chemical composition, P_H; T is temperature; C, chemical addition; x, clarity of pulp that is added; and τ residence time. The index 0 refers to a reference value and y_0 is the corresponding value of y. The parameters a to e are determined from the process operation. Equation (13.19) is a linearized equation around the operating point.

Amrehn[72] uses a simple dynamic model for batch polymerization reactors. From a change in temperature with time the amount of cooling required is predicted with a model. With the aid of partial mass balances the monomer conversion is predicted.

CHAPTER 14

Optimization

Chapter 12 dealt with a certain class of optimization problems, namely those where the process has to operate close to constraints. Some simple techniques were introduced for control of these systems. In this chapter attention will be focused on optimization from an economic point of view. The objective function for optimization will be defined and certain processes reviewed in which optimization might improve overall plant operation. It is not the intent of this chapter to present several optimization techniques but just to provide an introduction to the subject. However, after studying this chapter the reader will be able to better understand some of the opportunities for improved control and optimization in his or her own working environment and become familiar with some of the terminology used in optimizing control.

DYNAMIC OPTIMIZATION

Many production processes are operated dynamically. Air preheaters for steel furnaces, for example, operate on a cycle basis with hot furnace gas and combustion air. There is an optimal time for both cycles. Another example is the scheduling of reheating furnaces between the steel production process and the hot plate roller. Each steelplate has to be heated to a certain temperature, requiring a certain residence time in the furnace. The problem is to control the supply and discharge of steelplates in such a way that they have the right temperature at the right moment and are ready for transportation to the rolling mill. In the chemical industry some absorption processes are operated in a cyclic fashion with flows at minimum and maximum values for certain periods of time. Dynamic optimization often requires complicated and lengthy calculations which have to be carried out off-line.

In modern control theory and for simple applications in the process industry, the process is usually described by a set of differential equations, with the objective function being the integral of deviations over time. Optimization means the determination of control as a function of time; in some cases this can be translated into optimal feedback or feed-forward control with the aid of curve fitting.

A well-known example is the linear quadratic problem, where the quadratic objective function is optimized for a set of linear differential equations. This objective function, however, does not always represent the economic objective function of the process; hence, dynamic optimization does not always give the complete answer for optimal process operation.

STATIC OPTIMIZATION

In static optimization the process is described by an algebraic set of equations and an economic objective function, indicating how the profit of the process depends on process conditions, values of raw materials and products, and so forth.

In the process industries there are usually three different optimization problems: allocation, scheduling, and operating conditions. Allocation has to do with the best distribution of a flow or liquid over parallel processing units; scheduling has to do with the right time for maintenance (cleaning of process equipment as heat exchangers, decoking of furnaces, etc.) and can be combined with the scheduling problem as indicated earlier. Optimization of process conditions means the determination of optimal values for process units, such as optimal reflux flow and pressure for a distillation column. The economic objective function can depend on the independent variables in different ways. There can be a monotonous increasing or decreasing relationship, in which minimum and maximum constraints play a role. There can also be a parabolic relationship where the optimum lies on the "top of the hill." Whether this optimum is reached or not depends on the existing constraints. For the determination of the optimal operating point(s) many techniques are available.[73] Figure 14.1 presents an overview of hierarchy in control and shows where process optimization fits in. It is the highest level of control, after a basic and advanced level of control have been firmly established.

THE OBJECTIVE FUNCTION

In optimization an objective function and constraints have to be added to the process model. The general form of the objective function is:

$$J = \Sigma c_p F_p - \Sigma c_f F_f - \Sigma c_u F_u - c_o \qquad (14.1)$$

in which
- J = "profit rate" ($/time unit)
- c_p = value of product flow ($/kg)
- c_f = value of feed ($/kg)
- c_u = value of utilities, such as steam and cooling water ($/kg)
- F_p = product flow (kg/time unit)

F_f = feed flow (kg/time unit)
F_u = utility "flow" (kg/time unit)
c_o = fixed costs ($/time unit)

The term "profit" has a special meaning. It is not part of the profit, but a measure for the efficiency of the process operation. The term "value" is not strictly equal to the purchase and selling price, but is related to higher levels in the control hierarchy. The constraints are imposed on the manipulated input variables, the output variables, the process equipment and conditions, and on the range in which the model is valid.

Optimization means the maximization of the objective function within the constraints and according to the functional relationships of the model. Mathematically this means nonlinear programming for static models.

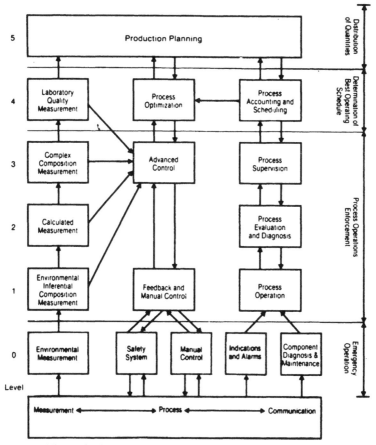

Figure 14.1. Hierarchy in control.

ADVISORY CONTROL

A first step of optimization involves the process operators (on-line open loop). The computer calculates optimal values for the remaining independent process variables and informs the operators how to steward toward these targets.

It is advisable to start off with this before "closing" the loop. Incompleteness or bugs in the supervisory control programs can now be identified and corrected without causing disastrous effects in process operation. This approach also gives the operators the opportunity to familiarize themselves with the new approach. Advisory control has been applied e.g., by Duyfjes and Van der Grinten[43] in a distillation process where parallel units and recycling via a chemical reactor contributed to the complexity.

CLOSED LOOP CONTROL

If one has enough confidence in the system, one can "close the loop." The general structure under which the processes now will be operated is shown in Figure 14.2. The supervisory optimal control program has to be designed in such a way that it enables the operator to take over the control functions. This is applied in an oxo-synthesis process,[44] where a comparison is made between conventional process operation, advisory control, and closed loop supervisory control.

It appeared that the process operators were unable to make the frequent but small corrections that were made by the on-line computer. Human beings are apparently more inclined to make one large adjustment in one process variable, rather than a number of small adjustments in a few variables. Also in other applications the experience was that a combination of small adjustments in all manipulated variables is not easily accepted. This is of special importance if this combination explicitly avoids the passing of a critical constraint.

OPTIMIZATION IN INDUSTRY

An excellent article on optimization in the refining industry is by J.P. Kennedy.[74] This article will be followed closely in defining some of the objectives for various process operations in the refining industry.

Crude Distillation

Crude units are large energy consumers in refineries, and they are the main piece of equipment for separation of components for further processing. Opti-

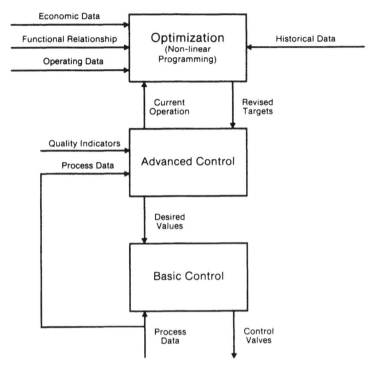

Figure 14.2.   Closed loop optimal control.

mal crude unit control can minimize energy costs and maximize the amount of distillate product. A technique that is increasingly accepted is inferential composition control, which maintains component properties during crude changes. The boiling range of each product is computed from measured temperatures and flows, and the calculation is reset with off-line quality measurements. This approach is well accepted and good results have been obtained.[75,76] Richalet et al.[77] implemented optimal control of a distillation process by modeling the process by its impulse response and identifying the model parameters on-line. The process is optimized within the constraints and an idea is given about the optimization incentives. Also, Baxley[78] optimized a distillation process, by using a Fenske-Underwood-Gilliland model for the determination of optimal operation within given constraints. Other examples are the optimization of train of distillation columns[79] and the determination of the optimal reflux ratio for a propylene unit.[80]

In-Line Blending

The purpose of in-line blending is to combine intermediate products from different plants or units into salable products. The primary objective is holding a set of ratios between blends. The higher level task is to set optimal blend ratios and to make changes to produce product within specification with minimal giveaway. The volatility and octane number both have to be controlled. The latter, however, is hard to control at the blender; the only easy way to change it is with the reformer severity.

The function of the blender control strategy is to provide an early warning that volatility or octane number is out of balance and to transmit this back to the refinery. This information will then be used to determine fractionation cut points, reformer severity, hydrocracker conversion per pass, and feedrate to the units.

Catalytic Reforming

The objective of the system is to maintain octane number and volatility targets at their desired values. Octane number and volatility interact at the blender, hence a consistent reformate will make blending easier and will reduce giveaway.

A catalytic reformer control system will maintain a computed octane number by manipulating furnace outlet temperatures within the constraints.[81] Volatility is controlled by using an inferential composition for the reformate product to set the debutanizer operation. The other controls that are used are the hydrogen/feed ratio and feed analysis to determine the best operating conditions.[82]

Hydrocrackers

The primary function of the control system is to deliver constant property components for sale and blending and to serve as feedstocks to other plants. The control package[75,83,84] consists of:

- The properties of the hydrocracker streams are held constant using inferential composition controls and/or volatility analyzers.

- The reactors are stabilized by controlling the predicted conversion per pass.

- The material inventory is controlled by adjusting the conversion in the reactors or feedrate.

- Runaway is detected and halted automatically.

Ethylene Plant

Ethylene plants are particulary suitable for optimization because of the large volume of feedstocks processed. Minor changes can therefore result in large financial gain or loss. The approach to the optimization of an entire ethylene plant is given in Sourander et al.[85] The plant model used, schematically given in

Figure 14.3, is complex. The objective of the furnace control system is to stabilize the furnace effluent composition for a given feed to yield the maximum gross plant margin without violating the furnace's physical constraints. In developing the control strategy, the interactions have to be considered and decoupled to reach the overall objectives. Stabilizing the yields means stabilizing the severity and selectivity at a constant feed composition. Severity is defined as the C_2/C_3 ratio and selectivity as the C_1/C_2 ratio. The furnace control hierarchy is organized as shown in Figure 14.4. Each of the coils has a feed flow controller and steam flow controller. The outlet temperature controller of target coil C adjusts the fuel gas pressures simultaneously by changing the setpoints of the pressure controllers. The heat load to individual coils can be balanced by adjusting the setting of the fuel gas pressure ratio controllers.

Cracking severity is controlled by manipulating coil outlet temperatures (COTs). The COTs for three coils are controlled by manipulating the coil feed rates. The total hydrocarbon throughput of the heater is maintained by adjusting the target coil feed rate. Selectivity is controlled by maintaining a certain

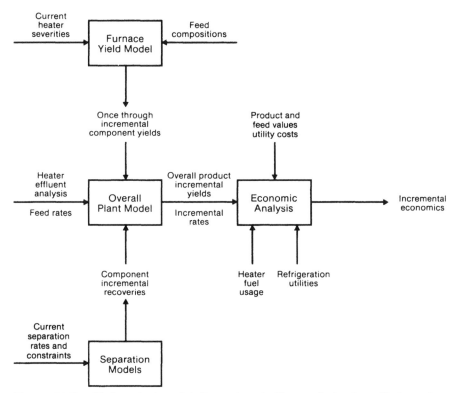

Figure 14.3.   Olefins plant model. Reproduced with permission from *Hydrocarbon Processing*, June 1984.

steam to hydrocarbon ratio. The variables for plant optimization are the furnace feedrates and severity targets for different plant feeds. The cracked gas compressor suction pressure is not set by the optimizer since a minimum suction pressure was found to be optimal. Constraints considered in the optimization include projected furnace load and maximum severity limits for each feed type as well as ethane recycle limits. Separation constraints are also included in the optimization.

The operator can choose two objective functions: one leads to maximization of the ethylene production against plant constraints; the other is an economic function that includes feedstock costs, product values, furnace fuel consumption, and steam consumption. The most important feature of the plant model is that it predicts incremental moves. If absolute moves are predicted, moves in the wrong direction can be made during transients in plant operation. The

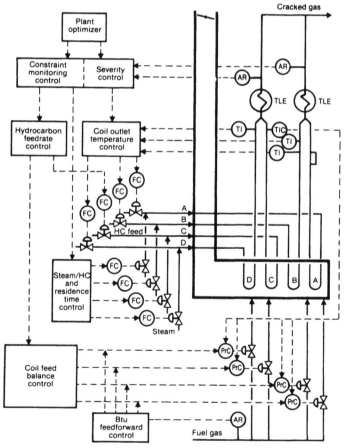

Figure 14.4. Furnace control scheme. Reproduced with permission from *Hydrocarbon Processing*, June 1984.

furnace yield model is used to calculate incremental composition and yield values. Incremental recoveries can be directly translated in incremental economic value.

The optimization problem is solved by linearizing the models and using linear programming. Typical execution time of the optimizer is once every four hours. The following savings in terms of changes of operating conditions are reported:

- with severity 2% to 10% higher than before, furnace run lengths have increased by 30%.

- coil outlet temperature changes have been reduced from 3.5°C to 1°C.

- furnace throughput has increased 4%.

PARALLEL UNITS

Sometimes it happens that a feed is distributed over a number of parallel units or that a number of parallel units have to produce a certain product. The optimization can then be simplified by application of Lagrange multipliers. Unit i has the following "profit rate":

$$J_i = f_i(u_i) \tag{14.2}$$

The total flow that is used or produced is:

$$F_i = \Sigma F_i(u_i) \tag{14.3}$$

The objective function is now defined as follows:

$$J = \sum_i f_i(u_i) + \lambda[F - \sum_i F_i(u_i)] \tag{14.4}$$

in which λ = Lagrange multiplier
                 u_i = manipulated variable for unit i
                 J_i = profit rate for unit i
                 F_i = flow for unit i

Differentiation of J with respect to u gives:

$$\frac{dJ}{du_i} = \frac{df_i}{du_i} - \frac{dF_i}{du_i} = 0 \tag{14.5}$$

Differentiation with respect to λ gives equation (14.3) again. Equation (14.5) only comprises λ and the variables of unit i. The optimization is now desegregated in optimizations per unit, which produces considerable time savings. Through iteration the value of λ has to be determined in such a way that equation (14.3) is satisfied. One should not forget that this procedure assumes a real maximum. In a mathematical notation for each unit:

$$\frac{d^2 J}{du_i^2} < 0 \qquad\qquad (14.6)$$

If this is not satisfied, the procedure results in the most unfavorable solution.

An example is the production of steam by a number of identical boilers. The efficiency of a boiler usually increases with load. This means that it is advantageous to run all boilers at maximum load except one that can cope with fluctuations in the steam consumption. A totally different picture exists in the case of parallel identical compressors, where the efficiency reduces by increase of load. In this case it is optimal to run all compressors at the same load. Other examples of applications can be found in power plants[86] and distillation columns.[80] It is obvious that when the parallel processes are not identical, one starts with the most efficient process. When this is running at capacity, the next process is started up until the least efficient process is finally started up.

CONSTRAINT OPTIMIZATION

For some processes an increase in input will result in decreased profit and decrease in additional product. This gives an optimal operating point in which the incremental value of the production is equal to the incremental cost of, for example, the utility consumption. In reality this optimum can not always be reached because one or more constraints will limit, for example, the consumption of water, electricity, or steam. Then the optimal value lies on the constraint. In complicated production processes there are usually several degrees of freedom for optimization and there are several constraints, of which some may be critical; it is obvious to use a control algorithm that controls the process within these constraints. This will result in a control system with two levels (Figure 14.5). The upper level (optimization) adjusts the setpoints of noncritical variables and selects the control scheme.

A difficult case of a critical constraint exists when many variables determine the distance to the constraint, especially if different lags are involved as well. An example is the gas compressor in a catalytic cracking unit (Figure 14.6). The load of the compressor is determined by the conditions in the reactor and the distillation column. The calculation of the constraint variable must be carried out with exceptional care. An interesting application can be found in

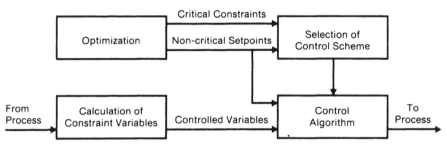

Figure 14.5.   Global constraint control.

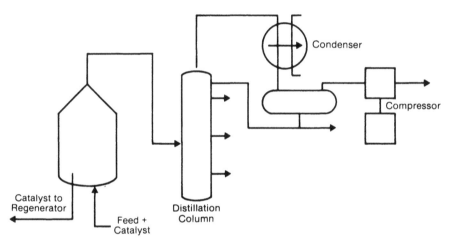

Figure 14.6.   Diagram of a catalytic cracking installation.

Laspe,[87] in which the optimization of an ethylene production unit and a fluid-ized catalytic cracking unit are discussed. In the catalytic cracking unit ten degrees of freedom were available: the feed, recycle feed, regenerator air flow, the stripping steam flow, the reactor steam flow, the slurry recycle flow, the amount of regenerated and used catalyst, the outlet air temperature of the preheater, and the system pressure. The constraints for the system were: maxi-mum air compressor flow, compressor constraint for the distillation column, pressure drops across valves, maximum regeneration temperature, maximum preheat temperature, maximum values for all flows, and a maximum system pressure. An example of the profit function is given in Figure 14.7. Savings due to the optimization were about 175 K$/year.

Hydrocarbon Processing[88] issued a special report, "Advanced Process Con-trol Handbook." Advanced control and optimization of various processes are presented with process flow diagrams and brief software descriptions. The

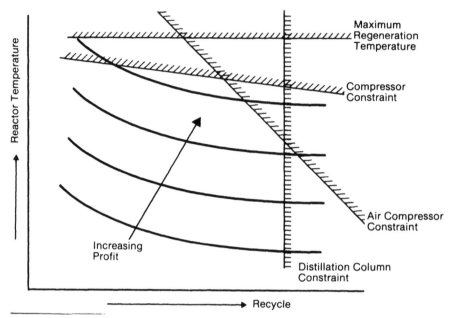

Figure 14.7.   Operating diagram for a catalytic cracking unit.

report gives a good impression about which commercial control packages are currently on the market.

CHAPTER 15

Self-Tuning and Adaptive Controllers

Successful implementation of a control scheme usually involves a number of steps:

- collection of process information

- identification and characterization of a process model (and disturbance if required)

- design of a controller based on selected criteria

- implementation of the controller

Usually one does not go through all these steps, since most processes can be controlled by simple PI or PID feedback control. However, in situations where the process has a major dead time or where the process model changes, it may be necessary to perform the design process and repeat it regularly.

A process model could change due to a number of reasons: decaying catalyst, fouling of equipment, change in process throughput, and so forth. Self-tuning control will perform all controller design and implementation steps online. The self-tuning controller design was developed by Aström and Wittenmark,[89] while Clarke and Gawthorp[90] developed the self-tuning constrained controller, which will be discussed in this chapter. The literature on self-tuning controllers is very extensive. Excellent review papers are given, however, by Aström,[91] who gives 164 references; Seborg et al.[92] with 247 references; and Unbehauen,[93] who gives 330 references. A paper giving suggestions for practical implementation of self-tuners is the one by Aström and Wittenmark.[94]

SELF-TUNING CONTROL

A self-tuning controller has the structure as shown in Figure 15.1. The self-tuning controller uses a recursive on-line parameter estimator, which makes

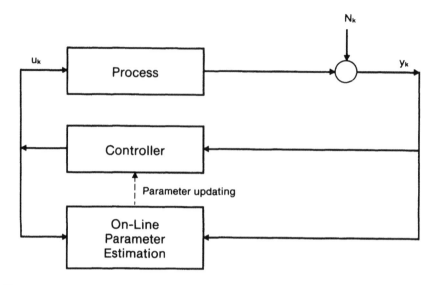

Figure 15.1.   Self-tuning controller.

use of the input and output information of the process in order to tune the controller parameters. The controller structure will depend on the process model and disturbance model. In this chapter the nomenclature will be used as adopted by Harris et al.[95] Assume that we have the following process model:

$$y_{k+f} = \frac{\omega(z^{-1})u_k}{\delta(z^{-1})} + N_{k+f} \tag{15.1}$$

in which $\omega(z^{-1}), \delta(z^{-1})$ are polynomials of the backward difference operator z^{-1}
f is the process delay time plus one additional period for the sample and hold
y is the output deviation from its target value
u is the control input

The disturbance N_{k+f} is modeled as follows:

$$N_{k+f} = \frac{\theta(z^{-1})}{\phi(z^{-1})\nabla} a_{k+f} \tag{15.2}$$

in which

$\theta(z^{-1}), \phi(z^{-1})$ are polynomials of the backward difference operator z^{-1}
a_{k+f} is white noise
$\nabla = 1 - z^{-1}$

Disturbance model factorization gives:

$$N_{k+f} = \frac{\theta(z^{-1})}{\phi(z^{-1})\nabla} a_{k+f} = \psi(z^{-1})a_{k+f} + \frac{K(z^{-1})}{\phi(z^{-1})\nabla} a_k \qquad (15.3)$$

where $\psi(z^{-1})$ is a polynomial of order f – 1.

The first term on the right contains only the effect of future disturbances and represents the minimum forecast error at time k, whereas the second term represents the forecast.

Substituting equation (15.3) into (15.1) gives:

$$[y_{k+f} - \psi(z^{-1})a_{k+f}] = \frac{\omega(z^{-1})}{\delta(z^{-1})} u_k + \frac{K(z^{-1})}{\phi(z^{-1})\nabla} a_k \qquad (15.4)$$

Substituting:

$$a_k = \frac{\phi(z^{-1})\nabla}{\theta(z^{-1})} \left\{ y_k - \frac{\omega(z^{-1})z^{-f}}{\delta(z^{-1})} u_k \right\} \qquad (15.5)$$

into equation (15.4) and collecting terms yields:

$$[y_{k+f} - \psi(z^{-1})a_{k+f}] = \frac{K(z^{-1})}{\theta(z^{-1})} y_k + \frac{\omega(z^{-1})}{\delta(z^{-1})} \left[\frac{\theta(z^{-1}) - K(z^{-1})z^{-f}}{\theta(z^{-1})} \right] u_k \qquad (15.6)$$

Now using the identity:

$$\theta(z^{-1}) - K(z^{-1})z^{-f} = \psi(z^{-1})\phi(z^{-1})\nabla \qquad (15.7)$$

which comes from equation (15.3), and multiplying by $\delta(z^{-1}) \theta(z^{-1})$ gives:

$$\delta(z^{-1})\theta(z^{-1})[y_{k+f} - \psi(z^{-1})a_{k+f}] \qquad (15.8)$$
$$= \delta(z^{-1})K(z^{-1})y_k + \omega(z^{-1})\psi(z^{-1})\phi(z^{-1})\nabla u_k$$

which can be written as:

$$\delta(z^{-1})\theta(z^{-1})[y_{k+f} - \epsilon_{k+f}] = \alpha(z^{-1})y_k + \beta^*(z^{-1})\nabla u_k \qquad (15.9)$$

where

$$\alpha(z^{-1}) = \alpha_0 + \alpha_1 z^{-1} + \ldots + \alpha_m z^{-m} \qquad (15.10)$$

$$\beta^*(z^{-1}) = \beta_0^* + \beta_1^* z^{-1} + \ldots + \beta^* z^{-l} \qquad (15.11)$$

$$\epsilon_{k+f} = f - \text{step ahead forecast error } \psi(z^{-1})a_{k+f}$$

If the parameters of equation (15.9) were known, the control action

$$\nabla u_k = -\frac{\alpha(z^{-1})}{\beta^*(z^{-1})}\, y_k = -\frac{\delta(z^{-1})}{\omega(z^{-1})}\, \frac{K(z^{-1})}{\phi(z^{-1})\psi(z^{-1})}\, y_k \tag{15.12}$$

would minimize $E\{y^2_{k+f}\}$, the expectation of the variance of the deviation of the process output from its setpoint. The control law of equation (15.12) is called a minimum variance controller. The term $\delta(z^{-1})/\omega(z^{-1})$ is the inverse of the process model without dead time, the term $K(z^{-1})/\phi(z^{-1})$ is the transfer function for the optimal disturbance predictor [see equation (15.3)], and the term $\psi(z^{-1})$ is the optimal compensator for f periods of delay.

PROCESS AND DISTURBANCE MODEL

Many processes in the chemical industry can be described by a first order and dead time model:

$$G(s) = \frac{Ke^{-s\theta}}{1 + s\tau} \tag{15.13}$$

Depending on the sampling time of the control scheme the discrete model may have an integer or real number of delay intervals.
The discrete representation of equation (15.13) can be given as:

$$G(z^{-1}) = \frac{(\omega_0 - \omega_1 z^{-1})z^{-f}}{(1 - \delta_1 z^{-1})} \tag{15.14}$$

in which $\delta_1 = e^{-\Delta t/\tau}$
$\quad\quad\quad f$ = number of whole periods of delay (the integer part of $\theta/\Delta t$) plus one additional period of delay for the sample and hold
$\quad\quad\quad c$ = fractional period of delay, $\theta/\Delta t - f$
$\quad\quad\quad \omega_0 = K(1 - \delta_1^{1-c})$
$\quad\quad\quad \omega_1 = K(\delta_1 - \delta_1^{1-c})$
$\quad\quad\quad \Delta t$ = sampling interval

Example

Assume the following transfer function:

$$G(s) = 0.5\,\frac{e^{-6s}}{1 + 12s}$$

For $\Delta t = 3$ we get:

$$G(z^{-1}) = \frac{0.1\ z^{-3}}{1 - 0.8\ z^{-1}}$$

The disturbance model can take different forms.[96,97] In case of noisy drifting disturbances, as shown in Figure 15.2 the following model would be a good representation:

$$N_k = \frac{(1 - xz^{-1})}{(1 - z^{-1})}\ a_k, \quad x > 0 \qquad (15.15)$$

The denominator term $(1-z^{-1})$ will force integral action in the controller, which is required because of the drifting character of the disturbances.

The model of equation (15.15) could also be used for randomly occurring setpoint changes. For step disturbances this model could be used with a value of x close to or equal to zero.

In case of smooth disturbances, e.g., load disturbances, it would be better to use a different disturbance model:

$$N_k = \frac{1}{(1 - xz^{-1})(1 - z^{-1})}\ a_k \qquad (15.16)$$

This type of disturbance is shown in Figure 15.3. For $x = 0$ the disturbance model reduces to what is commonly called a random walk. In the disturbance model of equation (15.15) a larger value of x, $0 \leq x \leq 1$, will represent a drifting process disturbance with an increasing amount of superimposed measurement noise.

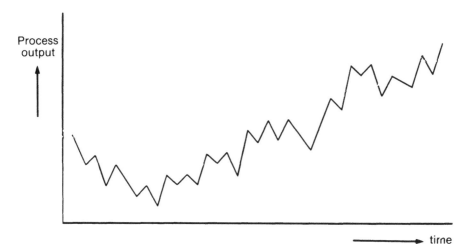

Figure 15.2.   Noisy drifting disturbance.

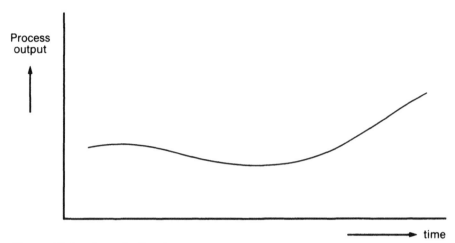

Figure 15.3. Smooth disturbance.

Let us now combine the process and disturbance model and work out equation (15.9).

For the process model we will use:

$$G(z^{-1}) = \frac{0.1 \ z^{-3}}{1 - 0.8 \ z^{-1}} \tag{15.17}$$

and for the disturbance model:

$$N_k = \frac{(1 - 0.1 \ z^{-1})}{(1 - z^{-1})} \ a_k \tag{15.18}$$

Factorization of equation (15.18) using equation (15.3) gives:

$$N_{k+3} = [1 + (1 - 0.1)z^{-1} + (1 - 0.1)z^{-2}]a_{k+3} + \frac{(1 - 0.1)z^{-3}}{(1 - z^{-1})} \ a_k \tag{15.19}$$

From comparison with the equations given previously we see that:

$$\omega(z^{-1}) = 0.1$$
$$f = 3$$
$$\delta(z^{-1}) = 1 - 0.8 \ z^{-1}$$
$$\theta(z^{-1}) = 1 - 0.1 \ z^{-1}$$
$$\psi(z^{-1}) = 1 + 0.9 \ z^{-1} + 0.9 \ z^{-2}$$
$$K(z^{-1}) = 0.9 \ z^{-3}$$
$$\phi(z^{-1}) = 1$$

and therefore equation (15.9) would become:

$$(1 - 0.8\ z^{-1})(1 - 0.1\ z^{-1})[y_{k+3} - \epsilon_{k+3}] = \qquad (15.20)$$

$$(1 - 0.8\ z^{-1}) \cdot 0.9\ z^{-3}y_k + 0.1 \cdot [1 + 0.9\ z^{-1} + 0.9\ z^{-2}]\nabla u_k$$

Equation (15.9) is the model for process and disturbances. In case $\delta(z^{-1})\theta(z^{-1}) = 1.0$ it can be simplified to:

$$y_{k+f} = \alpha(z^{-1})y_k + \beta^*(z^{-1})\nabla u_k + \epsilon_{k+f} \qquad (15.21)$$

In case $\delta(z^{-1})\theta(z^{-1}) \neq 1.0$, Harris et al.[95] have shown that equation (15.21) can be written as:

$$y_{k+f} = \alpha(z^{-1})y_k + \beta^*(z^{-1})\nabla u_k +$$

$$\mu_1[\alpha(z^{-1})y_{k-1} - \frac{\hat{\alpha}(z^{-1})_{k-1}}{\hat{\beta}(z^{-1})_{k-1}} \beta(z^{-1})y_{k-1}] + \qquad (15.22)$$

$$\mu_2[\alpha(z^{-1})y_{k-2} - \frac{\hat{\alpha}(z^{-1})_{k-2}}{\hat{\beta}(z^{-1})_{k-2}} \beta(z^{-1})y_{k-2}] + \ldots \quad \epsilon_{k+f}$$

The best estimates for α and β^* are $\hat{\alpha}$ and $\hat{\beta}$. Upon parameter convergence the third, fourth, and other terms on the RHS of equation (15.22) tend to go to zero and therefore the process and disturbance model is well approximated by equation (15.21).

CONSTRAINED INPUT CONTROL

Minimum variance control sometimes results in excessively large variations in the manipulated variable. It is common practice to calculate the control law which minimizes the variance of y subject to a constraint on the variance of ∇u_k, i.e., that minimizes:

$$J = E\{(y_{k+f})^2 + \zeta'(\nabla u_k)^2\} \qquad (15.23)$$

It can be shown that this is equivalent to minimizing:[90]

$$J^* = E\{y_{k+f} + \zeta\nabla u_k\}^2 = E\{\phi_{k+f}\}^2 \qquad (15.24)$$

with $\zeta = \zeta'/\omega_0$.

Using the definition of ϕ_{k+f} as in equation (15.24), equation (15.9) may be expressed:

$$\phi_{k+f} - \epsilon_{k+f} = \alpha(z^{-1})y_k + [\beta^*(z^{-1}) + \zeta\delta(z^{-1})\theta(z^{-1})]\nabla u_k \qquad (15.25)$$

where $\alpha(z^{-1})$ and $\beta^*(z^{-1})$ are as defined before and ζ is the penalty for excessive "valve movement." Equation (15.26) can be written as:

$$\phi_{k+f} = \alpha(z^{-1})y_k + \beta(z^{-1})\nabla u_k + \epsilon_{k+f} \qquad (15.26)$$

in analogy with equation (15.21). The order of $\beta(z^{-1})$ may be higher than the order $\beta^*(z^{-1})$. It should be noted that the sign of ζ should be that of ω_0.

Example

For a first order system with a process delay of two periods the model in analogy with equation (15.26) – see also equation (15.20) – would be:

$$\phi_{k+3} = \hat{\alpha}_0 y_k + \hat{\alpha}_1 y_{k-1} + \hat{\beta}_0 \nabla u_k + \hat{\beta}_1 \nabla u_{k-1} + \hat{\beta}_2 \nabla u_{k-2} \qquad (15.27)$$

where the hat refers to the estimated value.
In terms of the present time:

$$\phi_k = \hat{\alpha}_0 y_{k-3} + \hat{\alpha}_1 y_{k-4} + \hat{\beta}_0 \nabla u_{k-3} + \hat{\beta}_1 \nabla u_{k-4} + \hat{\beta}_2 \nabla u_{k-5} \qquad (15.28)$$

with

$$\phi_k = y_k + \zeta \nabla u_{k-3} \qquad (15.29)$$

If the dead time of the process is not exactly known, it is better to overestimate it than to underestimate it. In case of underestimation the self-tuning controller becomes easily unstable.

The controller structure is derived from the condition $\phi_{k+3} = 0$, therefore:

$$\nabla u_k = -\frac{1}{\hat{\beta}_0}[\hat{\alpha}_0 y_k + \hat{\alpha}_1 y_{k-1} + \hat{\beta}_1 \nabla u_{k-1} + \hat{\beta}_2 \nabla u_{k-2}] \qquad (15.30)$$

The self-tuning controller structure based on (15.28) and (15.30) is shown in Figure 15.4. One of the coefficients in the model could turn out to be very small, in which case the corresponding term could be deleted.

A different version of the constrained controller was developed by Ydstie et al.[98] A minimum variance controller tries to eliminate the deviation of the process variable from setpoint in one time step and can therefore cause excessively large process input changes. Clarke and Gawthrop[90] introduce a penalty factor $\zeta(\nabla u_k)^2$ which will restrict process input changes.

Ydstie et al.[98] proposed to eliminate the deviation of the process variable

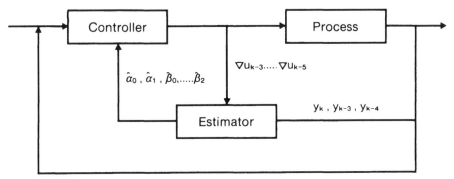

Figure 15.4. Constrained self-tuning controller structure.

from setpoint in a number of steps rather than one; the control strategy is therefore called extended horizon self-tuning controller (EHSTC).

In a practical application Ydstie's approach and Clarke and Gawthrop's approach will lead to similar results: a limitation of excessive changes in process input. An added advantage of the EHSTC is that by extending the control horizon, the controller can "look beyond" time delays, thus making it unnecessary to predict the time delay and making it possible to deal with variable time delays.

SELECTION OF THE SAMPLING INTERVAL

The sampling interval will depend on the time constant of the process and disturbances. Assume that in a particular case the process time constants are τ = 6 min and θ = 20 min and that for the disturbances τ = 10 min.

The frequency of a control tag in this case would typically be three to five minutes. As a rule of thumb:

$$\Delta t = (^1/_3 \ldots \ ^1/_5)\tau_{min} \qquad (15.31)$$

where $\tau_{min} = \text{MIN}(\tau_{process}, \tau_{disturbance})$.

If the tag execution interval Δt becomes too small, however, $\theta/\Delta t$ will become too large. And each period of delay will add one extra term to the model since:

$$\psi(z^{-1}) = 1 + (1 - \theta)z^{-1} + (1 - \theta)z^{-2} + \ldots + (1 - \theta)z^{-f+1} \qquad (15.32)$$

with f equal to $(\theta/\Delta t + 1)$.

Thus, for very small Δt the model will become too complex and convergence of model parameters may take long or never be achieved. Typically the number of terms in equation (15.32) should be limited to three or four, which means a value of the process time constant equal to two or three times the tag execution interval.

As a general rule the process model should be as simple as possible. It is better to start with a model that is too simple and increase the number of terms later on, than to start with a model that has too many terms. The other important thing that should be mentioned is that the higher the number of model parameters, the more excitation will be required to produce accurate parameter estimates.

SOME SPECIAL CASES

PI control

The controller structure of equation (15.30) can be generalized to:

$$(\beta_0 + \beta_1 z^{-1} + ... + \beta_\ell z^{-\ell}) \ \nabla u_k = (\alpha_0 + \alpha_1 z^{-1} + ... + \alpha_m z^{-m})(y_k - y_{sp}) \qquad (15.33)$$

comes in because of non-stationary disturbances

dead time compensation

$\ell \geq f-1$, f = periods of delay = process delay + 1

In case there is no dead time, $\beta_1 \ ... \ \beta_\ell$ are zero. The value of β_0 can be arbitrarily chosen. Control of a first order system (two alphas) gives:

$$\nabla u_k = \alpha_0 e_k + \alpha_1 e_{k-1} \qquad (15.34)$$

with

$$e_k = y_k - y_{sp} \qquad (15.35)$$

The algorithm for a PI controller can be given as:

$$\nabla u_k = K_c(e_k - e_{k-1} + \frac{\Delta t}{\tau_i} e_k) \qquad (15.36)$$

which gives for the gain and integral action:

$$K_c = -\alpha_1 \qquad (15.37)$$

$$\tau_i = -\Delta t \ \frac{\alpha_1}{\alpha_0 + \alpha_1} \qquad (15.38)$$

PID Control

The self-tuning PID controller can be obtained when one more term is added to the model. The control law then becomes:

$$\nabla u_k = \alpha_0 e_k + \alpha_1 e_{k-1} + \alpha_2 e_{k-2} \tag{15.39}$$

The PID algorithm can be given as:

$$\nabla u_k = K_c [e_k - e_{k-1} + \frac{\Delta t}{\tau_i} e_k + \frac{\tau_d}{\Delta t} (e_k - 2e_{k-1} + e_{k-2})] \tag{15.40}$$

Comparing equations (15.39) and (15.40) gives:

$$K_c = -\alpha_1 + 2\alpha_2 \tag{15.41}$$

$$\tau_i = -\Delta t \, \frac{\alpha_1 + 2\alpha_2}{\alpha_0 + \alpha_1 + \alpha_2} \tag{15.42}$$

$$\tau_d = -\Delta t \, \frac{\alpha_2}{\alpha_1 + 2\alpha_2} \tag{15.43}$$

LEAST SQUARES ESTIMATION

The model shown in equation (15.26) can be written as:

$$\hat{\phi}_k = x^T_{k-f} \hat{\theta} + \epsilon_k \tag{15.44}$$

in which
$$x^T_{k-f} = (y_{k-f}, y_{k-f-1} \cdots y_{k-f-m}, \nabla u_{k-f}, \nabla u_{k-f-1} \cdots \nabla u_{k-f-\ell}) \tag{15.45}$$
$$\hat{\theta}^T = (\hat{\alpha}_0, \hat{\alpha}_1 \cdots \hat{\alpha}_m, \hat{\beta}_0, \hat{\beta}_1 \cdots \hat{\beta}_\ell) \tag{15.46}$$
$$\phi_k = y_k + \zeta \nabla u_{k-f} \tag{15.47}$$
$$f = \text{process delay} + 1$$
$$\zeta = \text{penalty on valve movement}$$

The hat refers to the estimated value. The parameters of equation (15.44) are determined such that the least squares criterion

$$J(\theta) = \sum_{s=1}^{k} \lambda^{k-s} \epsilon^2(s) \qquad (15.48)$$

is minimized; epsilon (ϵ) is the error between estimated model parameters and their actual value. Lambda (λ) is a discounting factor, which reduces the impact of past data on the current estimates. Effectively $1/(1-\lambda)$ data points are included in the estimation. The value of $1/(1-\lambda)$ is called the effective window length. Thus if $\lambda = 0.95$, the effective window length is 20; effectively the estimator works only with the past 20 values. Lambda is usually in the range $0.95 \leq \lambda \leq 1$.

The solution may be expressed as:[96,99]

$$\begin{aligned}\hat{\underline{\theta}}_k &= \hat{\underline{\theta}}_{k-1} + \underline{K}_k \, [\phi_k - \hat{\phi}_k] \\ &= \hat{\underline{\theta}}_{k-1} + \underline{K}_k \, [y_k + \zeta \nabla u_{k-f} - \underline{x}^T_{k-f} \hat{\underline{\theta}}] \end{aligned} \qquad (15.49)$$

$$\underline{K}_k = \frac{\underline{\underline{P}}_{k-1} \underline{x}_{k-f}}{(\lambda + \underline{x}^T_{k-f} \underline{\underline{P}}_{k-1} \underline{x}_{k-f})} \qquad (15.50)$$

$$\underline{\underline{P}}_k = \frac{1}{\lambda} \left[\underline{\underline{P}}_{k-1} - \frac{\underline{\underline{P}}_{k-1} \underline{x}_{k-f} \underline{x}^T_{k-f} \underline{\underline{P}}_{k-1}}{\lambda + \underline{x}^T_{k-f} \underline{\underline{P}}_{k-1} \underline{x}_{k-f}} \right] \qquad (15.51)$$

$$= \frac{1}{\lambda} \left[\underline{\underline{P}}_{k-1} - \underline{K}_k \underline{x}^T_{k-f} \underline{\underline{P}}_{k-1} \right]$$

The notation is that a single bar represents a vector and a double bar a matrix. $\underline{\underline{P}}_k$ is the symmetric matrix $(\underline{x}^T \underline{x})^{-1}$ at time k. This matrix is called the covariance matrix, and it is a positive semi-definite matrix that is used to update the variance of each parameter (the diagonal elements) as well as the covariances of the parameters with each other (off-diagonal elements).

The vector $\underline{K}$ is used to update the adjustments of each parameter. In case of no delay f = 1.

Initial estimates for $\hat{\underline{\theta}}(o)$ and $\underline{P}(o)$ are needed to start the calculation. Wittenmark and Aström[94] suggest that $\underline{P}(o)$ be chosen as:

$$10 \, \alpha \, I \leq P(0) \leq 100 \, \alpha \, I \qquad (15.52)$$

where α is the variance of the output variable y and I is the unit matrix.

Lambda (λ) usually starts at 0.95 and could go exponentially, for example, to 0.99 using

$$\lambda_k = 0.95 \lambda_{k-1} + 0.05 * 0.99 \qquad (15.53)$$

When new information comes in the form of increased process excitation, λ should be reset to 0.95 in order to facilitate parameter adjustment.

SELF-TUNING FEED-FORWARD CONTROL

Assume that we have a process as shown in Figure 15.5.

In addition to unobserved disturbances N_k, fluctuations in the process output y_k may be due to change in a process variable w_k, which can be measured but not manipulated.

If there were no unobserved disturbances N_k and no control action was taken, the process output would equal w_k^*, where:

$$w^*_k = \frac{\omega^*(z^{-1})}{\delta^*(z^{-1})} \, w_{k-f^*} = \frac{\omega^*_0 + \omega^*_1 z^{-1} + \dots + \omega^*_s z^{-s}}{1 + \delta^*_1 z^{-1} + \dots + \delta^*_r z^{-r}} \, w_{k-f^*} \qquad (15.54)$$

w_k is referred to as a feed-forward variable. In the case where $f^* \geq f$ (in other words, the manipulated variable u_k can compensate for the measured disturbance before it reaches the output y_k), the controller minimizing $E\{y^2_{k+f}\}$ is given by:

$$\nabla u_k = - \frac{\delta(z^{-1})}{\omega(z^{-1})} \left[\frac{\omega^*(z^{-1})}{\delta^*(z^{-1})} \, \nabla w_{k-j} + \frac{K(z^{-1})}{\phi(z^{-1})\psi(z^{-1})} \, y_k \right] \qquad (15.55)$$

in which

$$j = f^* - f \qquad (15.56)$$

When we compare equation (15.55) to equation (15.12), we see that the control law contains an extra term which accounts for the measured disturbance w.

If $f^* > f$, then ∇w_{k-j} has not yet occurred at time k and the controller of equation (15.55) is not physically realizable. A minimum variance forecast of $\nabla \hat{w}_{k-j/k}$ is then made and substituted in place of $[\omega^*(z^{-1}) \nabla w_{k-j}]/[\delta^*(z^{-1})]$.[97]

If the parameters of equation (15.55) were unknown, one might estimate them from a model of the form:

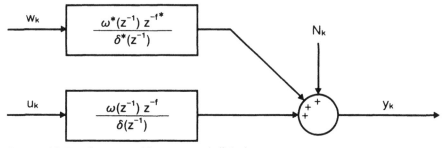

Figure 15.5. Process with measured disturbance.

$$y_{k+f} = \alpha(z^{-1})y_k + \beta(z^{-1})\nabla u_k + \nu(z^{-1})\nabla w_{k-j} \qquad (15.57)$$

where

$$\left.\begin{array}{l} \alpha(z^{-1}) = \alpha_0 + \alpha_1 z^{-1} + \ldots + \alpha_m z^{-m} \\[4pt] \beta(z^{-1}) = \beta_0 + \beta_1 z^{-1} + \ldots + \beta_l z^{-l} \\[4pt] \nu(z^{-1}) = \nu_0 + \nu_1 z^{-1} + \ldots + \nu_n z^{-n} \end{array}\right\} \qquad (15.58)$$

and use them in the control law

$$\nabla u_k = - \left[\frac{\hat{\nu}(z^{-1})}{\hat{\beta}(z^{-1})} \nabla w_{k-j} + \frac{\hat{\alpha}(z^{-1})}{\hat{\beta}(z^{-1})} y_k \right] \qquad (15.59)$$

as if they were exactly known.

ESTIMATOR INITIALIZATION AND CONVERGENCE

A precaution that should be observed is the choice of $\hat{\beta}_0$ (see equation [15.30]). Its initial value should reflect the action of the controller. For a normally acting controller $\hat{\beta}_0$ will be positive. The value of the other parameters can be set to zero or any good guess.

If the regulator is put into service for the first time, then the estimator could be started five to ten intervals before the controller is activated. Otherwise a good starting guess would be the previously converged values.

One of the biggest problems encountered with the self-tuning controller is that of round-off error. When initiating the estimator, it is normal to set the diagonal elements of the covariance matrix to a large positive value and the off-diagonal elements to zero. This is equivalent to saying that the value of the parameters (α's and β's) have been set to a number we have little confidence in. However, this may lead to round-off error and ill-conditioning. The starting values of the diagonal elements of the P-matrix should therefore be chosen carefully.

It is important to note that the values of the diagonal elements of the covariance matrix should always be positive. Negative values indicate a problem with the estimator.

After the self-tuning algorithm has been on for some time, the parameter estimates will no longer change. Once the estimates have converged, the self-tuner has been tuned and the estimator part of the algorithm can be turned off.

APPLICATION TO A PROCESS WITHOUT DELAY (f = 1)

Assume that a process can be described by the following difference equation:

$$y_{k+1} = 1.7y_k - 0.7y_{k-1} + 0.6u_k - 0.4u_{k-1} - 0.2u_{k-2} + a_{k+1} - 0.7a_k \quad (15.60)$$

in which a_k is white noise.

In order to control this process, assume the following model:

$$y_{k+1} = \hat{\alpha}_0 y_k + \hat{\alpha}_1 y_{k-1} + \hat{\beta}_0 \nabla u_k \quad (15.61)$$

which can be written as:

$$\phi_{k+1} = \underline{x}^T_k \hat{\underline{\theta}} = y_{k+1} + \zeta \nabla u_k \quad (15.62)$$

in which

$$\left. \begin{array}{l} \underline{x}^T_k = (y_k, y_{k-1}, \nabla u_k) \\ \hat{\underline{\theta}}^T = (\hat{\alpha}_0, \hat{\alpha}_1, \hat{\beta}_0) \end{array} \right\} \quad (15.63)$$

From equations (15.50) and (15.51) it can be seen that the term $\underline{P}_{k-1}\underline{x}_{k-1}$ appears several times. Therefore this expression is usually evaluated first.

Note that the P-matrix is a 3×3 matrix since three model parameters have to be estimated. It is defined as:

$$\underline{P}_{k-1} = \begin{pmatrix} P_{11} & P_{12} & P_{31} \\ P_{12} & P_{22} & P_{32} \\ P_{13} & P_{23} & P_{33} \end{pmatrix}_{k-1} \quad (15.64)$$

Using equation (15.63), $\underline{x}_{k-1}$ can be written as:

$$\underline{x}^T_{k-1} = (y_k, y_{k-1}, \nabla u_k)_{k=k-1} \quad (15.65)$$

The product $\underline{P}_{k-1}\underline{x}_{k-1}$ now becomes:

$$\begin{pmatrix} P_{11}y_k + P_{12}y_{k-1} + P_{13}\nabla u_k \\ P_{12}y_k + P_{22}y_{k-1} + P_{23}\nabla u_k \\ P_{13}y_k + P_{23}y_{k-1} + P_{33}\nabla u_k \end{pmatrix}_{k=k-1} = \begin{pmatrix} SUM1 \\ SUM2 \\ SUM3 \end{pmatrix}_{k=k-1} \quad (15.66)$$

The expression $\lambda_{k-1} + \underline{x}^T_{k-1}\underline{P}_{k-1}\underline{x}_{k-1}$ therefore becomes:

$$(y_k, y_{k-1}, \nabla u_k) \begin{pmatrix} SUM1 \\ SUM4 \\ SUM6 \end{pmatrix}_{k=k-1} + \lambda_{k-1} =$$

$$[y_k \cdot SUM1 + y_{k-1} \cdot SUM2 + \nabla u_k \cdot SUM3 + \lambda]_{k=k-1} = SUM_{k-1} \quad (15.67)$$

This gives the following expression for K:

$$\underline{K}_k = \begin{pmatrix} K1 \\ K2 \\ K3 \end{pmatrix}_k = \begin{pmatrix} SUM1 \\ SUM2 \\ SUM3 \end{pmatrix}_{k-1} /SUM_{k-1} \qquad (15.68)$$

from which:

$$K1_k = SUM1_{k-1}/SUM_{k-1}, \text{ etc.} \qquad (15.69)$$

The expression for P_k can be found when is $x^T_{k-1} \underline{P}_{k-1}$ is evaluated first. It is not difficult to see that this becomes:

$$x^T_{k-1}\underline{P}_{k-1} = (SUM1, SUM2, SUM3)_{k=k-1} \qquad (15.70)$$

Therefore the following expression for P is obtained:

$$P_{11,k} = \frac{1}{\lambda_{k-1}}\left[P_{11} - \frac{SUM1 \cdot SUM1}{SUM} \right]_{k-1} \qquad (15.71)$$

$$P_{12,k} = \frac{1}{\lambda_{k-1}}\left[P_{12} - \frac{SUM1 \cdot SUM2}{SUM} \right]_{k-1}, \text{ etc} \qquad (15.72)$$

Equations (15.64) to (15.72) can be written conveniently in vector-matrix notation; however, to fully grasp the mathematical manipulations it is advisable to write them out at least once.

The control law can be derived from equation (15.61) by making $y_{k+1} = 0$:

$$\nabla u_k = -\frac{1}{\beta_0}[\hat{\alpha}_0 y_k + \hat{\alpha}_1 y_{k-1}] \qquad (15.73)$$

The setpoint of the controlled system (15.60) was changed stepwise between +2 and –2. Control action was calculated using $\lambda = 0.99$ and $\varsigma = 1.0$.

Figure 15.6 shows the step response for the following initial set of parameter estimates: $\hat{\alpha}_0 = 1.0$, $\hat{\alpha}_1 = 0.3$ and $\hat{\beta}_0 = 1.6$.

Figure 15.7 shows the convergence of parameter $\hat{\beta}_0$, and Figure 15.8 shows the trace (equal to the sum of diagonal elements of the P matrix). It can be seen that $\hat{\beta}_0$ (as well as $\hat{\alpha}_0$ and $\hat{\alpha}_1$) converges quickly to a new value.

APPLICATION TO A PROCESS WITH TWO PERIODS OF DELAY (f = 3)

Assume that the process can now be described by:

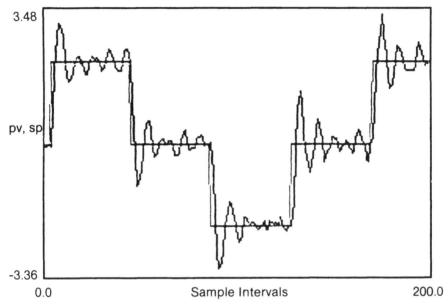

Figure 15.6.   Setpoint and process output for first order system without delay.

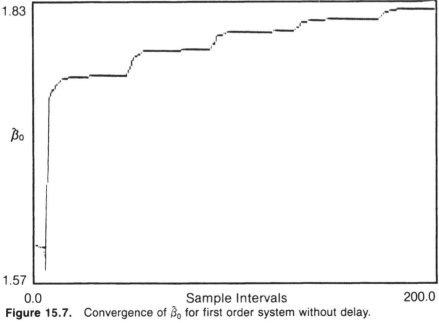

Figure 15.7.   Convergence of $\hat{\beta}_0$ for first order system without delay.

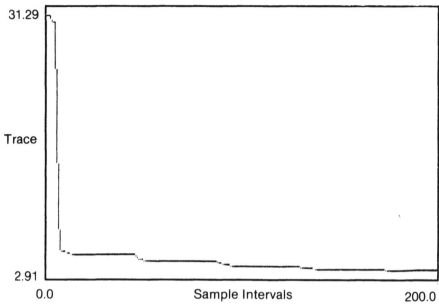

Figure 15.8. Trace as a function of time.

$$y_{k+1} = 1.7y_k - 0.7y_{k-1} + 0.6u_{k-2} - 0.6u_{k-3} + a_{k+1} - 0.7a_k \qquad (15.74)$$

Assume a model of the following form:

$$\phi_{k+3} = \hat{\alpha}_0 y_k + \hat{\alpha}_1 y_{k-1} + \hat{\beta}_0 \nabla u_k + \hat{\beta}_1 \nabla u_{k-1} + \hat{\beta}_2 \nabla u_{k-2} \qquad (15.75)$$

or

$$\phi_k = \hat{\alpha}_0 y_{k-3} + \hat{\alpha}_1 y_{k-4} + \hat{\beta}_0 \nabla u_{k-3} + \hat{\beta}_1 \nabla u_{k-4} + \hat{\beta}_2 \nabla u_{k-5}$$

Define:

$$\left. \begin{array}{l} x^T_{k-3} = (y_{k-3}, \, y_{k-4}, \, \nabla u_{k-3}, \, \nabla u_{k-4}, \, \nabla u_{k-5}) \\ \\ \theta^T = (\hat{\alpha}_0, \, \hat{\alpha}_1, \, \hat{\beta}_0, \, \hat{\beta}_1, \, \hat{\beta}_2) \end{array} \right\} \qquad (15.76)$$

hence the model can be written as:

$$\left. \begin{array}{l} \phi_k = x^T_{k-3}\theta \\ \\ \text{with} \quad \phi_k = y_k + \zeta \nabla u_{k-3} \end{array} \right\} \qquad (15.77)$$

The control law follows from $k + 3 = 0$, therefore:

$$\nabla u_k = -\frac{1}{\hat{\beta}_0}[\hat{\alpha}_0 y_k + \hat{\alpha}_1 y_{k-1} + \hat{\beta}_1 \nabla u_{k-1} + \hat{\beta}_2 \nabla u_{k-2}] \tag{15.78}$$

The estimation is done by:

$$\theta_k = \theta_{k-1} + K_k[\phi_k - \hat{\phi}_k]$$
$$= \theta_{k-1} + K_k[y_k + \zeta\nabla u_{k-3} - x^T_{k-3}\theta_{k-1}] \tag{15.79}$$

The gain K_k is computed from:

$$K_k = \frac{P_{k-1}x_{k-3}}{(\lambda + x^T_{k-3}P_{k-1}x_{k-3})} \tag{15.80}$$

and the covariance matrix P_k from:

$$P_k = \frac{1}{\lambda}[P_{k-1} - K_k x^T_{k-3}P_{k-1}] \tag{15.81}$$

P_{k-1} is as defined in equation (15.64).
$P_{k-1}\, x_{k-3}$ can be written as (compare equation 15.66):

$$\begin{pmatrix} P_{11}y_{k-2} + P_{12}y_{k-3} + P_{13}\nabla u_{k-2} + P_{14}\nabla u_{k-3} + P_{15}\nabla u_{k-4} \\ P_{12}y_{k-2} + P_{22}y_{k-3} + P_{23}\nabla u_{k-2} + P_{24}\nabla u_{k-3} + P_{25}\nabla u_{k-4} \\ P_{13}y_{k-2} + P_{23}y_{k-3} + P_{33}\nabla u_{k-2} + P_{34}\nabla u_{k-3} + P_{35}\nabla u_{k-4} \\ P_{14}y_{k-2} + P_{24}y_{k-3} + P_{34}\nabla u_{k-2} + P_{44}\nabla u_{k-3} + P_{45}\nabla u_{k-4} \\ P_{15}y_{k-2} + P_{25}y_{k-3} + P_{35}\nabla u_{k-2} + P_{45}\nabla u_{k-3} + P_{55}\nabla u_{k-4} \end{pmatrix}_{k=k-1} = \begin{pmatrix} SUM1 \\ SUM2 \\ SUM3 \\ SUM4 \\ SUM5 \end{pmatrix}_{k=k-1} \tag{15.82}$$

The product $x^T_{k-3}\, P_{k-1}\, x_{k-3}$ added to λ_{k-1} becomes:

$$(y_{k-3},\ y_{k-4},\ \nabla u_{k-3},\ \nabla u_{k-4},\ \nabla u_{k-5}) \begin{pmatrix} SUM1 \\ SUM2 \\ SUM3 \\ SUM4 \\ SUM5 \end{pmatrix}_{k=k-1} + \lambda_{k-1} =$$

$$\tag{15.83}$$

$$y_{k-3} \cdot SUM1_{k-1} + y_{k-4} \cdot SUM2_{k-1} + \ldots + \nabla u_{k-5} \cdot SUM5_{k-1} + \lambda_{k-1} = SUM_{k-1}$$

The control law in this case is derived from equation (15.75) by making $\hat{\phi}_{k-3} = 0$:

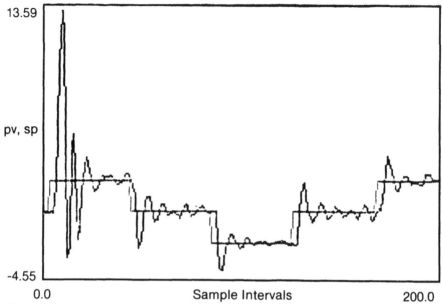

Figure 15.9. Setpoint and process output for first order system with delay (f = 3).

$$\nabla u_k = -\frac{1}{\hat{\beta}_0}[\hat{\alpha}_0 y_k + \hat{\alpha}_1 y_{k-1} + \hat{\beta}_1 \nabla u_{k-1} + \hat{\beta}_2 \nabla u_{k-2}] \qquad (15.84)$$

The first two terms in the right-hand side of this equation can be compared to the two terms of a PI controller; the other terms contain the history of past input moves and therefore compensate for the dead time.

Figure 15.9 shows the response of the self-tuning controller for $\lambda = 0.99$ and $\zeta = 1.0$ and the initial parameter vector $\theta(0) = (1.8, -1.5, 2.0, 0.4, 1.1)$. Figure 15.9 shows that after stepchanges the controller tunes itself in. If the large change in process output is unacceptable, one could limit the change in control action, and, as the controller gets tuned in, gradually increase this limit.

The response is obviously strongly dependent on the initial estimates of the controller parameters. It can be said that the initial parameter estimation becomes more critical as the number of periods of delay increases. This is not difficult to comprehend. A first order system can be easily controlled by a PI controller even if it is not tuned too well. From experience we know that a system with dead time needs to be well tuned and the estimation of the number of delay values is very important.

Figure 15.10 shows the change of model parameter $\hat{\beta}_0$ and Figure 15.11 shows the trace of P. From the pattern of trace with time it can be seen that confidence in the parameter estimates is quickly established.

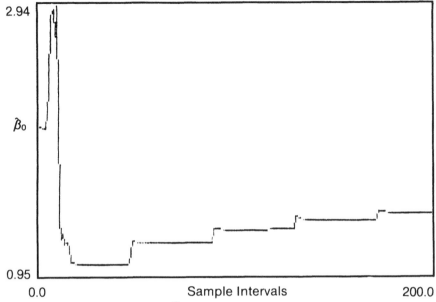

Figure 15.10. Convergence of $\hat{\beta}_0$.

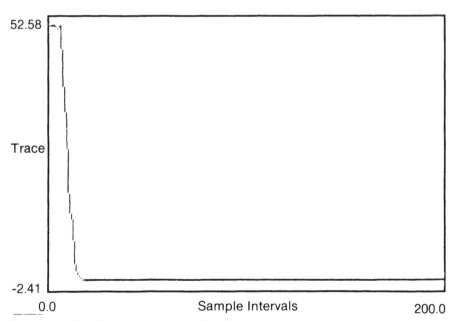

Figure 15.11. Trace for system with delay.

CHAPTER 16

State Observers

STATE ESTIMATION

An important question in the chemical industry is if it is possible to estimate important variables based on a limited number of measurements. To give an example: The concentration profile in a tubular reactor is a distributed parameter that can be measured usually only at the outlet; however, temperatures can be measured in more locations. Is it possible to estimate the entire concentration profile based on this limited number of measurements? Without a model, estimation of the concentration profile is possible using interpolation techniques; however, if we have a reasonably good model, it will be attractive to use it in order to improve the estimation. Obviously, without model and measurements an estimation of the concentration profile is impossible. A large number of estimation methods that all start with a state model has been developed. This boils down to the representation of the dynamic behavior by a set of simultaneous first order differential equations:

$$\frac{dx}{dt} = F_c(x,u,t) \tag{16.1}$$

in which
- x = vector of state variables
- u = vector of input variables
- F_c = function indicating continuous relationship
- t = time

Usually it is not very complicated to convert a process model into the form of equation (16.1). Energy, momentum, and continuity equations often have this form, and, in case of distributed processes, it is a matter of discretization with respect to the geometric coordinates. Also, algebraic relationships have to be substituted into the differential equations. Sometimes the differential equations have derivatives of more than one variable; in that case one has to separate these derivatives in the left-hand side of the equation.

The state model can be completed by measurement equations, in general form:

$$y = g_c (x,u,t) \tag{16.2}$$

in which y is the vector of output variables (of the measurement system) and where x, u, and t are as defined before.

Note the difference between measured variables, which are output variables of the process, and output variables of the measurement system, which are available for control, for example. The situation is shown in Figure 16.1.

Usually the number of output variables is less than the number of state variables. This is the reason for the problem definition: How do we estimate the state x based on a limited number of measurements? If there are no unknown variations, in theory the state can be estimated with a decreasing error. In that case we speak of an *observer*. However, if there are unknown (stochastic) variations, the estimation is always inaccurate. In that case we speak of a *filter*. In the latter case model equations (16.1) and (16.2) have to be modified to:

$$\frac{dx}{dt} = F_c(x,u,w,t) \tag{16.3}$$

and

$$y = g_c (x,u,v_y,t) \tag{16.4}$$

in which   w = the vector of unknown variations in the process, or process
noise
v_y = the vector of unknown variations in the measurement, or measurement noise

For linear systems, Luenberger defined the first observer in 1964 and Kalman and Bucy the first filter in 1960. Usually we speak of a Luenberger

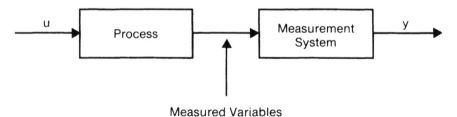

Measured Variables

Figure 16.1.   Process and measurement system.

observer and Kalman filter. In later years all kinds of improvements were made; however, applications in the chemical industry have been limited.

Since the observer and filter usually will be implemented on a process computer, we shall work with discrete models and the discrete version of an observer and filter.

Equations (16.1) to (16.4) then become:

$$x(k + 1) = F[x(k),u(k),k] \qquad (16.5)$$
$$y(k + 1) = g[x(k),u(k),k] \qquad (16.6)$$

and for the stochastic case:

$$x(k + 1) = F[x(k),u(k),w(k),k] \qquad (16.7)$$
$$y(k + 1) = g[x(k),u(k),v_y(k+1),k] \qquad (16.8)$$

The measurement equations implicitly include a certain measurement lag, which means that the state on time k has an impact on the measurement on time k + 1. It is also possible to formulate the problem without measurement lag, but this will lead to different results. Note that F is different from the function F_c in equations (16.1) and (16.3).

Further information on state estimation is found in Kwakernaak and Sivan,[100] Sage and Melsa,[101] and Gelb;[102] a good introduction to observers is given by Luenberger.[68,103,104]

STRUCTURE OF AN OBSERVER

Next an observer will be developed for the system described by equations (16.5) and (16.6). It is assumed that u and y are available without measurement errors. If we know the initial state of the system x(0), then equations (16.5) and (16.6) can be used to calculate the state for any given time, and the problem is already solved. Usually, however, only an estimate, called x(0), is available. It is the task of the observer to make sure that the estimation process converges and improves and that, after some time, the right values are estimated (this is only possible if the system is observable).

The initial estimate x(0) is used for a *prediction of the state one step ahead* without using the measurements:

$$\check{x}(1/0) = F[\hat{x}(0),u(0),0] \qquad (16.9)$$
$$\check{y}(1/0) = g[\hat{x}(0),u(0),0] \qquad (16.10)$$

Note the $\hat{x}$ means the estimate of x; $\check{x}$ the prediction of x based on previous information; and $\check{x}(1/0)$ means prediction of x at time 1 based on information available or estimated at time 0. Figure 16.2 shows an information flow diagram of the prediction.

On time 1 a new measurement becomes available: y(1). The difference between this measurement and the prediction one step ahead is called "innovation."

$$\check{e}_y(1/0) = y(1) - \check{y}(1/0) \tag{16.11}$$

This innovation indicates that the prediction has to be corrected. If the innovation is relatively small (which is the target!), a linear correction is sufficient:

$$\hat{x}(1/1) = \check{x}(1/0) + G(1) \cdot \check{e}_y(1/0) \tag{16.12}$$

in which x(1/1) is the estimation of the state on time 1, based on prediction and the real measurement on time 1. This can be repeated for time 2 and so forth. For time k + 1 the equations are:

$$\check{x}(k+1/k) = F[\hat{x}(k/k), u(k), k] \tag{16.13}$$
$$\check{y}(k+1/k) = g[\hat{x}(k/k), u(k), k] \tag{16.14}$$
$$\check{e}_y(k+1/k) = y(k) - \check{y}(k+1/k) \tag{16.15}$$
$$\hat{x}(k+1/k+1) = \check{x}(k+1/k) + G(k+1) \cdot \check{e}_y(k+1/k) \tag{16.16}$$

The information flow diagram is now extended to Figure 16.3.

In the diagram one has to start at $\hat{x}(k/k)$. From the process and measurement model then follow $\check{x}(k+1/k)$ and $\check{y}(k+1/k)$. Using the actual measurement y(k+1) results in the innovation $\check{e}_y(k+1/k)$ being available and the correction $G(k+1)\,\check{e}_y(k+1/k)$, which then gives us a new estimation of the

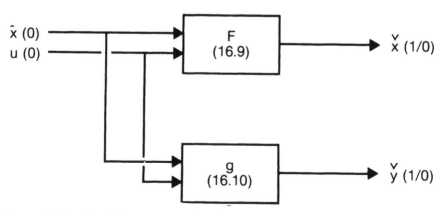

Figure 16.2.   Prediction one step ahead.

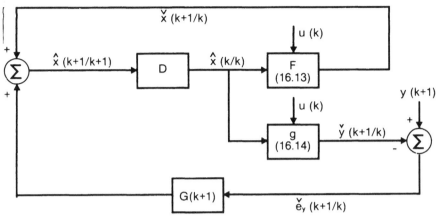

Figure 16.3.   Structure of an observer.

state $\hat{x}(k + 1/k + 1)$. The structure has been closed by introduction of the delay operator D, which indicates that one has to wait one discretization interval before executing the next step. The observer is a recursive algorithm, which can be used in real time. The only thing that remains to be done is the selection of the correction matrix $G(k + 1)$. This requires an expression for the estimation error.

DETERMINATION OF ESTIMATION ERRORS

The error in the estimation of the state on time $(k + 1)$ is given by:

$$\hat{e}(k + 1/k + 1) = x(k + 1) - \hat{x}(k + 1/k + 1) \tag{16.17}$$

Substitution of equation (16.16) gives:

$$\hat{e}(k + 1/k + 1) = x(k + 1) - \check{x}(k + 1/k) - G(k + 1) \cdot \check{e}_y(k + 1/k) \tag{16.18}$$

With the following expression for the prediction error

$$\check{e}(k + 1/k) = x(k + 1) - \check{x}(k + 1/k) \tag{16.19}$$

this becomes:

$$\hat{e}(k + 1/k + 1) = \check{e}(k + 1/k) - G(k + 1) \cdot \check{e}_y(k + 1/k) \tag{16.20}$$

Let's evaluate the prediction error $\check{e}(k + 1/k)$ first. With the aid of equations (16.5) and (16.13) this becomes:

$$\check{e}(k+1/k) = F[x(k),u(k),k] - F[\hat{x}(k/k),u(k),k] \qquad (16.21)$$

The first term in the RHS can be rewritten using a Taylor series development around $x(k/k)$:

$$F[x(k),u(k),k] = F[\hat{x}(k/k),u(k),k] + \phi(k)[x(k)-\hat{x}(k/k)] + \ldots \qquad (16.22)$$

in which:

$$\phi(k) = \left. \frac{\partial F}{\partial x} \right|_{x(k)=\hat{x}(k/k)} \qquad (16.23)$$

When ignoring higher order terms in equation (16.22), equation (16.21) becomes (with the aid of equations [16.22] and [16.17]):

$$\check{e}(k+1/k) = \phi(k) \cdot \hat{e}(k/k) \qquad (16.24)$$

Similarly, with the aid of equations (16.15), (16.14), and (16.6), we find for $\check{e}_y(k+1/k)$:

$$\check{e}_y(k+1/k) = g[x(k),u(k),k] - g[\hat{x}(k/k),u(k),k] \qquad (16.25)$$

When again we restrict ourselves to the first term in the Taylor series approximation, this becomes:

$$\check{e}(k+1/k) = H(k) \cdot \hat{e}(k/k) \qquad (16.26)$$

in which:

$$H(k) = \left. \frac{\partial g}{\partial x} \right|_{x(k)=\hat{x}(k/k)} \qquad (16.27)$$

Now we can evaluate the estimation error in equation (16.20) further. Substitution of equation (16.26) gives:

$$\hat{e}(k+1/k+1) = \check{e}(k+1/k) - G(k+1) \cdot H(k) \cdot \hat{e}(k+1/k) \qquad (16.28)$$

Substitution of equation (16.24) gives:

$$\hat{e}(k+1/k+1) = [\phi(k) - G(k+1) \cdot H(k)]\hat{e}(k/k) \qquad (16.29)$$

which gives the recurrent relationship we were looking for. The estimation error on time $k+1$ can now be determined from the estimation error on time k. Equations (16.28) and (16.29) are given in the flow diagram of Figure 16.4.

Note the similarity with Figure 16.3. The minus sign of the right addition point has moved to the left addition point.

SELECTION OF THE GAIN G

We still have to make a good choice for $G(k+1)$. A reasonable target is a fast reduction of the estimation error, which can be quite large for time 0. The matrix in the RHS of equation (16.29):

$$S(k+1) = [\phi(k)-G(k+1)\cdot H(k+1)] \qquad (16.30)$$

is of crucial importance in this process.

Repeated application of equation (16.29) gives (with equation [16.30]):

$$\hat{e}(k+1/k+1) = S(k+1)S(k) \ldots S(1)\hat{e}(0/0) \qquad (16.31)$$

which can be simplified when using the geometric average S:

$$\hat{e}(k+1/k+1) = \bar{S}^{(k+1)}\hat{e}(0/0) \qquad (16.32)$$

From matrix theory we know that a matrix S can be written in the form:

$$\bar{S} = T^{-1} JT \qquad (16.33)$$

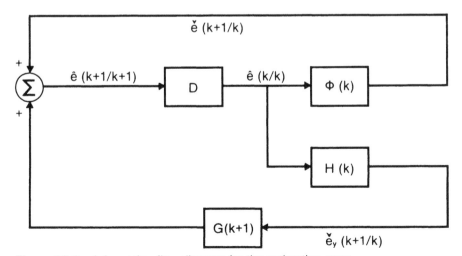

Figure 16.4.   Information flow diagram for the estimation error.

in which J is the Jordan canonical matrix.

If S can be diagonalized, then J is a diagonal matrix with eigenvalues on the main diagonal.

Then we have

$$\bar{S}^{k+1} = (T^{-1} JT) \ldots (T^{-1} JT) \ldots (T^{-1} JT) = T^{-1}J^{k+1}T \qquad (16.34)$$

in which J^{k+1} is the following matrix:

$$J^{k+1} = \begin{pmatrix} \lambda_1^{k+1} & & & & \\ & \lambda_2^{k+1} & & & \\ & & \cdots & & \\ & & & \cdots & \\ & & & & \lambda_n^{k+1} \end{pmatrix} \qquad (16.35)$$

Equation (16.32) can therefore only converge (for $k \rightarrow \infty$) if all eigenvalues are smaller than 1.

$$|\lambda_j| < 1 \text{ for } j = 1 \ldots n \qquad (16.36)$$

In order to have maximum convergence we can try to make all eigenvalues equal to zero. In that case the estimation error for a time-independent system reduces to zero in exactly n steps. That is a so-called deadbeat observer.

In general the system will not be time-independent (see equation 16.30). However, if we stay close to an average measurement, convergence is usually relatively fast.

ILLUSTRATION

To illustrate the selection of G consider a simple example. Around the average measurement, a system is described by:

$$\left. \begin{array}{l} x_1(k+1) = 0.9x_1(k) \\ x_2(k+1) = 0.9x_2(k) + x_1(k) \end{array} \right\} \qquad (16.37)$$

x_2 is measured, hence the observer is used to estimate x_1. Using equation (16.37) matrix ϕ is:

$$\phi = \begin{pmatrix} 0.9 & 0 \\ 1 & 0.9 \end{pmatrix} \qquad (16.38)$$

Matrix H is: $H = (0\ 1)$ (16.39)

hence equation (16.30) becomes:

$$S = \begin{pmatrix} 0.9 & 0 \\ 1 & 0.9 \end{pmatrix} - \begin{pmatrix} g_1 \\ g_2 \end{pmatrix} (0\ 1) = \begin{pmatrix} 0.9 & -g_1 \\ 1 & 0.9-g_2 \end{pmatrix}$$ (16.40)

The eigenvalues can be determined from:

$$\begin{vmatrix} 0.9-\lambda & -g_1 \\ 1 & 0.9-g_2-\lambda \end{vmatrix} = 0$$ (16.41)

or

$$\lambda^2 - \lambda (1.8-g_2) + 0.9 (0.9-g_2) + g_1 = 0$$ (16.42)

Both eigenvalues are zero if:

$$1.8 - g_2 = 0$$
$$0.81 - 0.9\ g_2 + g_1 = 0$$ (16.43)

from which gains $g_1 = 0.81$ and $g_2 = 1.8$ follow.

To determine the sensitivity for the operating point, 0.9 in equation (16.38) is replaced by 1.0. Using the calculated values for g, equation (16.42) now becomes:

$$\lambda^2 - 0.2\lambda + 0.01 = 0$$ (16.44)

or

$$\lambda_1 = \lambda_2 = 0.1$$

If we use 0.8 instead of 0.9 we find $\lambda_1 = \lambda_2 = -0.1$. Apparently the sensitivity for changes in the operating point is not undesirably large.

CHAPTER 17

Kalman Filtering

In the previous chapter state estimation in general was discussed; the Kalman filter was mentioned as state estimator for the stochastic case. As will be shown, the Kalman filter looks very much like the observer structure that was shown in the previous section.

An added feature is optimization, therefore the Kalman filter can be seen as an advanced application of the least squares method in dynamic estimation problems.

The fundamentals of Kalman filtering are given in Kalman,[105] Kalman and Bucy,[106] Jazwinski,[107] and Brown;[108] applications are given, amongst others, in Kipperissides et al.,[109] Jo and Bankoff,[110] Schuler,[111] and Canavas.[112]

PROCESS NOISE

The process model has already been introduced in the previous chapter (equations [16.6] and [16.7]). Equation (16.6) will be used instead of equation (16.8) and the measurement noise will be introduced later in a slightly different way than was done in equation (16.8).

For the time being the model is therefore:

$$x(k + 1) = F[x(k), u(k), w(k), k] \qquad (17.1)$$

$$y(k + 1) = g[x(k), u(k), k] \qquad (17.2)$$

in which $x(k)$ = state vector of the process
$u(k)$ = input vector of the process
$w(k)$ = process noise vector
$y(k)$ = output vector
k = time

The uncertainties in the process model can be translated into a contribution toward the process noise. We shall limit ourselves to the case where the average value of $w(k)$ is zero and where the signal shows relatively small stochastic variations.

The generality of the model is not violated if w(k) is a vector of a number of time-uncorrelated random variations (so-called discrete white noise). In mathematical expression:

$$E[w(k)] = 0 \quad \text{for all values of k} \tag{17.3}$$

$$E[w(k)w^T(j)] = 0 \quad \text{for } k \neq j \tag{17.4}$$

in which E is the expectation and "super T" denotes transpose. Equation (17.4) is the dyadic product of two vectors and can be written as follows:

$$\left.\begin{array}{llll} w_1(k)w_1(j) & w_1(k)w_2(j) & ..... & w_1(k)w_n(j) \\ w_2(k)w_1(j) & w_2(k)w_2(j) & ..... & w_2(k)w_n(j) \\ ..... & ..... & ..... & ..... \\ w_n(k)w_1(j) & w_n(k)w_2(j) & ..... & w_n(k)w_n(j) \end{array}\right\} \tag{17.5}$$

Note: If equation (17.4) is not satisfied, it is usually possible to develop a model that has discrete white noise with w_r as input vector and colored noise w as output vector (see Figure 17.1).

The noise model is then:

$$x_r(k + 1) = F_r[x_r(k), w_r(k), k] \tag{17.6}$$

$$w(k) = g_r[x_r(k), w_r(k), k] \tag{17.7}$$

in which $x_r(k)$ represents a state vector. Often it is possible to design a reasonable process noise model based on process knowledge. Equations (17.1), (17.6), and (17.7) can now be combined to:

$$\left.\begin{array}{l} x(k + 1) = F^0[x(k), x_r(k), w_r(k), k] \\ x_r(k + 1) = F_r[x_r(k), w_r(k), k] \end{array}\right\} \tag{17.8}$$

in which F^0 is the result of substitution of (17.7) into (17.1). If x and x_r together are treated as a new state vector, equation (17.8) now has exactly the same form as equation (17.1); hence, the original model description is maintained.

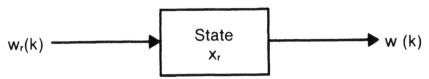

$w_r(k)$ →   State X_r →   w (k)

Figure 17.1. Modeling of noise.

STRUCTURE OF THE KALMAN FILTER

For further analysis it is assumed that the measurement noise is additive (compare to equation [17.3]):

$$y_m(k + 1) = y(k + 1) + v_y(k + 1) \qquad (17.9)$$

in which y = vector of measured variables (output variables)
y_m = vector of measured values

It is also assumed that the input variables are measured with a certain noise:

$$u_m = u(k) + v_u(k) \qquad (17.10)$$

We assume that v_y and v_u are discrete white noise signals. Figure 17.2 shows the information flow diagram.

Using an approach similar to that used for the observer in Chapter 16, we are going to make the best possible prediction of the state at time $k+1$, without using the measurements at $k+1$ (one step ahead prediction). Using equations (17.1) and (17.2) gives (compare to [17.3] and [17.4]):

$$\check{x}(k + 1/k) = F[\hat{x}(k/k), \hat{u}(k/k), 0, k] \qquad (17.11)$$

$$\check{y}(k + 1/k) = g[\hat{x}(k/k), \hat{u}(k/k), k] \qquad (17.12)$$

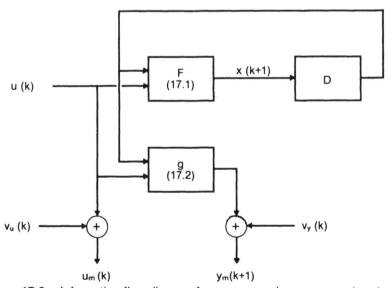

Figure 17.2. Information flow diagram for process and measurement system.

in which $\hat{x}$ means the estimate of x, $\check{x}$ the prediction of x, $\check{x}(k+1/k)$ means the prediction of x at time $k+1$ based on information available at time k.

In equations (17.11) and (17.12) zero has been substituted for the expected value of the process noise (see equation [17.3]).

The estimation $\hat{u}(k/k)$ follows from equation (17.10) with zero for the expected value of v_u:

$$\hat{u}(k/k) = u_m(k) \tag{17.13}$$

Substitution of (17.13) in equations (17.11) and (17.12) gives:

$$\check{x}(k + 1/k) = F[\hat{x}(k/k), u_m(k), 0, k] \tag{17.14}$$

$$\check{y}(k + 1/k) = g[\hat{x}(k/k), u_m(k), k] \tag{17.15}$$

Note: With the aid of equations (17.24), (17.25), (17.31), and (17.32) the previous two equations can also be written as:

$$\check{x}(k + 1/k) = \phi(k)\hat{x}(k/k) + \psi(k)u_m(k) \tag{17.14a}$$

$$\check{y}(k + 1/k) = H(k)\,\hat{x}(k/k) + \Omega(k)u_m(k) \tag{17.15a}$$

The prediction equations are shown in the information flow diagram of Figure 17.3.

At time $k+1$ a new measurement becomes available: $y_m(k+1)$.

The difference with the prediction is the innovation:

$$\check{e}_y(k + 1/k) = y_m(k + 1) - \check{y}(k + 1/k) \tag{17.16}$$

The next state estimate will be based on the previous prediction corrected by the calculated innovation (similar to equation [16.16]):

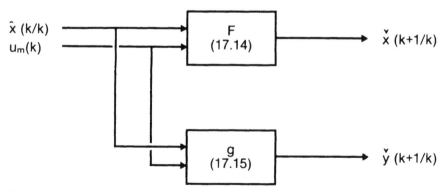

Figure 17.3.   Prediction one step ahead.

$$\hat{x}(k + 1/k + 1) = \check{x}(k + 1/k) + G(k + 1)\check{e}_y(k + 1/k) \qquad (17.17)$$

The information flow diagram can now be extended to Figure 17.4.

Note that this structure is similar to the structure of the observer shown in Figure 16.3. The only thing that remains is the selection of the gain $G(i + 1)$. For the Kalman filter the approach is totally different and the calculation is done via the least squares method. In order to illustrate this we first will determine the estimation errors and then the variances of the estimation errors.

DETERMINATION OF ESTIMATION ERRORS

The error in the estimation of the state at time $(k + 1)$ is given by:

$$\hat{e}(k + 1/k + 1) = x(k + 1) - \hat{x}(k + 1/k + 1) \qquad (17.18)$$

Substitution of equation (17.17) gives:

$$\hat{e}(k + 1/k + 1) = x(k + 1) - \check{x}(k + 1/k) - G(k + 1)\check{e}_y(k + 1/k) \quad (17.19)$$

With the expression for the prediction error:

$$\check{e}(k + 1/k) = x(k + 1) - \check{x}(k + 1/k) \qquad (17.20)$$

this becomes:

$$\hat{e}(k + 1/k + 1) = \check{e}(k + 1/k) - G(k + 1)\check{e}_y(k + 1/k) \qquad (17.21)$$

Let us evaluate the prediction error $\check{e}(k + 1/k)$ first. According to equations (17.1) and (17.14):

$$\check{e}(k + 1/k) = F[x(k),u(k),w(k),k] - F[\hat{x}(k/k),u_m(k),0,k] \qquad (17.22)$$

The first term in the RHS can be rewritten using a Taylor series development:

$$\begin{aligned}
F[x(k),u(k),w(k),k] = \ & F[\hat{x}(k/k),u_m(k),0,k] \\
& + \phi(k) [x(k) - \hat{x}(k/k)] \\
& + \psi(k) [u(k) - u_m(k)] \\
& + \Gamma(k) w(k) + \ldots..
\end{aligned} \qquad (17.23)$$

in which:

$$\phi(k) = \left. \frac{\partial F}{\partial x} \right|_{x(k) = \hat{x}(k/k)} \qquad (17.24)$$

$$\psi(k) = \frac{\partial F}{\partial u}\bigg|_{u(k)=u_m(k)} \tag{17.25}$$

$$\Gamma(k) = \frac{\partial F}{\partial w}\bigg|_{w(k)=0} \tag{17.26}$$

Ignoring higher order terms in equation (17.23) we can rewrite the equation with the aid of (17.18) and (17.10) as:

$$F[x(k),u(k),w(k),k] = F[\hat{x}(k/k),u_m(k),0,k] + \\ \phi(k)\hat{e}(k/k) - \psi(k)v_u(k) + \Gamma(k)w(k) \tag{17.27}$$

Substituted in equation (17.22) and ignoring the indices at time k gives:

$$\check{e}(k + 1/k) = \phi\hat{e} - \psi v_u + \Gamma w \tag{17.28}$$

Similarly, we find for $\check{e}_y(k+1/k)$ (see equation [17.16]):

$$\check{e}_y(k + 1/k) = y_m(k + 1) - \check{y}(k + 1/k) \\ = g[x(k),u(k),k] + v_y(k + 1) - g[\hat{x}(k/k),u_m(k),k] \tag{17.29}$$

With first order terms in the Taylor series development:

$$\check{e}_y(k + 1/k) = H\hat{e} - \Omega v_u + v_y(k + 1) \tag{17.30}$$

in which

$$H = \frac{\partial g}{\partial x}\bigg|_{x(k)=\hat{x}(k/k)} \tag{17.31}$$

$$\Omega = \frac{\partial g}{\partial u}\bigg|_{u(k)=u_m(k)} \tag{17.32}$$

Equations (17.28) and (17.30) substituted in (17.21) gives:

$$\hat{e}(k+1/k+1) = \phi\hat{e} - \psi v_u + \Gamma w - G(k+1)[H\hat{e} - \Omega v_u + v_y(k+1)] \tag{17.33}$$

or with index (k + 1) replaced by an asterisk:

$$\hat{e}^* = [\phi - G^*H]\hat{e} - [\psi - G^*\Omega]v_u + \Gamma w - G^*v^*_y \tag{17.34}$$

VARIANCE OF THE ESTIMATION ERRORS

We are interested in the variances of the estimation errors. In matrix notation it is common to put these variances in a covariance matrix. The variances are then located on the main diagonal; the covariances between the different variables are on the other locations.

$$P = \begin{pmatrix} E(\hat{e}_1{}^2) & E(\hat{e}_1\hat{e}_2) & ..... & E(\hat{e}_1\hat{e}_n) \\ E(\hat{e}_2\hat{e}_1) & E(\hat{e}_2\hat{e}_2) & ..... & E(\hat{e}_2\hat{e}_n) \\ ..... & ..... & ..... & ..... \\ E(\hat{e}_n\hat{e}_1) & E(\hat{e}_n\hat{e}_2) & ..... & E(\hat{e}_n{}^2) \end{pmatrix} \qquad (17.35)$$

Note that the covariance matrix is symmetric because $\hat{e}_i\hat{e}_j$ equals $\hat{e}_j\hat{e}_i$. The covariance matrix can therefore be written as the expectation of the dyadic product of e with itself:

$$P = E\left[\begin{pmatrix} e_1 \\ : \\ e_n \end{pmatrix} (e_1 \ ..... \ e_n) \right] = E[\hat{e}\hat{e}^T] \qquad (17.36)$$

Application to equation (17.34) gives:

$$\begin{aligned} P^* = E[\hat{e}^*\hat{e}^{*T}] = E\{&[(\phi - G^*H) -\hat{e} + \Gamma w \\ &- (\psi - G^*\Omega)v_u - G^*v^*{}_y][(\phi - G^*H)\hat{e} \\ &+ \Gamma w - (\psi - G^*\Omega)v_u - G^*v^*{}_y]^T\} \end{aligned} \qquad (17.37)$$

Rewriting of equation (17.37) with shifting of the expectation operator gives:

$$\begin{aligned} P^* = &(\phi - G^*H)E[\hat{e}\hat{e}^T](\phi - G^*H)^T + \Gamma E[ww^T]\Gamma^T + \\ &(\psi - G^*\Omega)E[v_u v_u{}^T](\psi - G^*\Omega)^T + G^*E[v^*{}_y v^*{}_y{}^T]G^{*T} \end{aligned} \qquad (17.38)$$

In this equation the terms with the covariance of $\hat{e}$ and w, $\hat{e}$ and v_u, $\hat{e}$ and $v^*{}_y$, w and v_u, w and $v^*{}_y$, and v_u and $v^*{}_y$ are independent of each other. Equation (17.38) in short notation is:

$$\begin{aligned} P^* = &(\phi - G^*H)P(\phi - G^*H)^T + \Gamma Q\Gamma^T + \\ &(\psi - G^*\Omega)N(\psi - G^*\Omega)^T + G^*R^*G^{*T} \end{aligned} \qquad (17.39)$$

in which

$$\left. \begin{aligned} P(k) &= E[\hat{e}\hat{e}^T] \\ Q(k) &= E[ww^T] \\ N(k) &= E[v_u v_u{}^T] \\ R(k + 1) &= R^* = E[v_y v_y{}^T] \end{aligned} \right\} \qquad (17.40)$$

MINIMIZATION OF THE VARIANCES OF THE ESTIMATION ERRORS

The next step is to minimize a weighted sum of the variances of the estimation errors:

$$J^* = \sum_{j=1}^{n} a^*_{jj} P^*_{jj} \tag{17.41}$$

in which a_{jj} are weighing coefficents.

In matrix notation equation (17.41) can be written as the trace (sum of the elements on the main diagonal) of matrix A^*P^* in which:

$$A^* = \begin{pmatrix} a^*_{11} & 0 & \cdots & \cdots & \cdots & 0 \\ 0 & a^*_{22} & 0 & \cdots & \cdots & 0 \\ \cdots & & 0 & \cdots & \cdots & \cdots \\ \cdots & \cdots & \cdots & \cdots & \cdots & \cdots \\ \cdots & \cdots & \cdots & \cdots & \cdots & \cdots \\ 0 & 0 & \cdots & \cdots & \cdots & a^*_{nn} \end{pmatrix} \tag{17.42}$$

$$J^* = \text{trace } A^*P^* = \sum_{j=1}^{n}\sum_{m=1}^{n} a^*_{jm} P^*_{mj} = \sum_{j=1}^{n} a^*_{jj} P^*_{jj} \tag{17.43}$$

With (17.39) equation (17.43) becomes:

$$J^* = \text{trace}[A^*(\phi - G^*H)P(\phi - G^*H)^\mathsf{T} + A^*\Gamma Q \Gamma^\mathsf{T} + \\ A^*(\psi - G^*\Omega)N(\psi - G^*\Omega)^\mathsf{T} + A^*G^*R^*G^{*\mathsf{T}}] \tag{17.44}$$

A necessary condition for an optimum is that the variance of J^* with respect to G^* is equal to zero.

$$\frac{\partial J^*}{\partial G^*} = 0 \tag{17.45}$$

When we apply the following rules for the differentiation of a trace:

$$\left. \begin{aligned} \frac{\partial}{\partial G} \text{ trace } [BGC] &= B^\mathsf{T}C^\mathsf{T} \\[2mm] \frac{\partial}{\partial G} \text{ trace } [BG^\mathsf{T}C] &= CB \end{aligned} \right\} \tag{17.46}$$

We can write equations (17.44) and (17.45) as:

$$\frac{\partial J^*}{\partial G^*} = -A^{*T}(\phi - G^*H)P^TH^T - A^*(\phi - G^*H)PH^T - A^{*T}(\psi - G^*\Omega)N^T\Omega^T \quad (17.47)$$

$$-A^*(\psi - G^*\Omega)N\Omega^T + A^{*T}G^*R^{*T} + A^*G^*R^* = 0$$

Since P, N, R*, and A are symmetric, equation (17.47) can be simplified to:

$$\frac{\partial J^*}{\partial G^*} = -2A^*(\phi - G^*H)PH^T - 2A^*(\psi - G^*\Omega)N\Omega^T + 2A^*G^*R^* = 0 \quad (17.48)$$

This equation can be rewritten as follows:

$$(\phi PH^T + \psi N\Omega^T) = G^*(HPH^T + \Omega N\Omega^T + R^*) \quad (17.49)$$

Solving this equation for G* gives:

$$G^* = (\phi PH^T + \psi N\Omega^T)(HPH^T + \Omega N\Omega^T + R^*)^{-1} \quad (17.50)$$

Here it is assumed that the last matrix may be inverted.

Finally we can use (17.50) to simplify equation (17.39). After some mathematical manipulations we get:

$$P^* = (\phi - G^*H)P\phi^T + (\psi - G^*\Omega)N\psi^T + \Gamma Q\Gamma^T \quad (17.51)$$

If so desired, we can also use the variances belonging to equations (17.28) and (17.30):

$$\tilde{P}(k + 1/k) = \phi P\phi^T + \psi N\psi^T + \Gamma Q\Gamma^T \quad (17.52)$$

$$\tilde{P}_y(k + 1/k) = HPH^T + \Omega N\Omega^T + R^* \quad (17.53)$$

Using these equations we can simplify equations (17.50) and (17.51):

$$G^* = (\phi PH^T + \psi N\Omega^T)[\tilde{P}_y(k + 1/k)]^{-1} \quad (17.54)$$

$$P^* = \tilde{P}(k + 1/k) - G^*(HP\phi^T + \Omega N\psi^T) \quad (17.55)$$

RESULTS

Let us summarize all equations. First of all the predictions one step ahead must be carried out:

$$\breve{x}(k + 1/k) = F[\hat{x}(k/k),u_m(k),0,k] \qquad (17.14)$$

$$\breve{y}(k + 1/k) = g[\hat{x}(k/k),u_m(k),k] \qquad (17.15)$$

Then follows the calculation of the variances of the predictions:

$$\breve{P}(k + 1/k) = \phi P \phi^T + \psi N \psi^T + \Gamma Q \Gamma^T \qquad (17.52)$$

$$\breve{P}_y(k + 1/k) = HPH^T + \Omega N \Omega^T + R^* \qquad (17.53)$$

Now follows the calculation of the gain calculation:

$$G(k + 1) = D[\breve{P}_y(k + 1/k)]^{-1} \qquad (17.54)$$

in which

$$D = \phi PH^T + \psi N \Omega^T \qquad (17.56)$$

Then the new state estimation:

$$\qquad\qquad\qquad\qquad\qquad\qquad\qquad\qquad\qquad (17.16)$$
$$\hat{x}(k + 1/k + 1) = \breve{x}(k + 1/k) + G(k + 1)[y_m(k + 1) - \breve{y}(k + 1/k)] \quad \text{and}$$
$$\qquad\qquad\qquad\qquad\qquad\qquad\qquad\qquad\qquad (17.17)$$

and the estimation of the new covariances:

$$P(k + 1/k + 1) = \breve{P}(k + 1/k) - G(k + 1)D^T \qquad (17.57)$$

or with (17.54)

$$P(k + 1/k + 1) = \breve{P}(k + 1/k) - D[\breve{P}_y(k + 1/k)]^{-1}D^T \qquad (17.58)$$

The last equation indicates that the new variances are smaller than the prediction variance one step ahead.

The total structure is shown in the flow diagram of Figure 17.5. Figure 17.4 is part of this structure.

It is obvious that the process provides the measured values $u_m(k)$ and $y_m(k + 1)$. The next step is to make estimates of the covariance matrices Q of the process noise w, N of the input measurement noise v_u, and R of the output measurement noise. Usually these values will be selected time independent although the Kalman filter is able to handle time-dependent values.

The algorithm has to be started with an estimated initial state x(0) and an estimated covariance matrix for this initial state.

Usually these initial estimates are not critical. If the process stays close to an average state, matrices ϕ, ψ, H, Ω, and Γ change little. In that case the calculation of variances can be done in an average operating point, and, in general, fast convergence of G to a final value is achieved. The entire lower

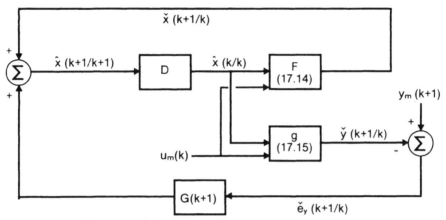

Figure 17.4.   Structure of the Kalman filter.

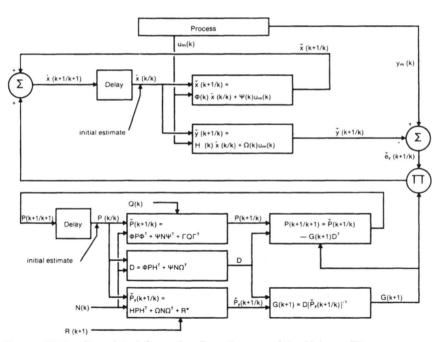

Figure 17.5.   Complete information flow diagram of the Kalman filter.

part of Figure 17.5 is only executed a limited number of times and the processing of measurements is done only via the upper part.

However, in real applications one is often not sure that the process model is correct. This uncertainty has to be translated into increased process noise.

Correct performance of the Kalman filter can be checked by observing the innovation (equation [17.6]). It can be shown that the innovation should have the character of white noise, i.e.,

$$E[\breve{e}_y(k + 1/k)\breve{e}_y^T(k/k - 1)] = 0 \qquad (17.59)$$

for the optimal gain G. Intuitively we had expected this: for an optimal filter a new measurement only brings "unexpected" information; if anything could be expected it would have been incorporated in the model.

COMPARISON TO THE LITERATURE

The derivation of the Kalman filter in many textbooks, e.g., Brown,[108] is usually somewhat different. In general the following simplifications are made:

1. No difference is made between prediction ($\breve{}$) and estimation ($\hat{}$), hence predicted parameters are treated as estimated parameters.

2. The input variables can be measured without noise: therefore equation (17.14) can be written as:

$$\hat{x}(k + 1/k) = F[\hat{x}(k/k),u(k),0,k]$$
$$= \phi(k)\, \hat{x}(k/k) + \psi(k)u(k) \qquad (17.14b)$$

3. No measurement lag is built into the measurement equation and also no dependence on u is assumed. Equation (17.15) can then be written as:

$$\hat{y}(k + 1/k) = g[\hat{x}(k + 1/k),k]$$
$$= H(k + 1)\, \hat{x}(k + 1/k) \qquad (17.15b)$$

Equations (17.22) and (17.23) can be rewritten using equation (17.14b). This leads to a somewhat modified equation (17.28):

$$\hat{e}(k + 1/k) = \phi\hat{e} + \Gamma w \qquad (17.28a)$$

Equation (17.29) is rewritten as:

$$\hat{e}_y(k + 1/k) = y_m(k + 1) - \hat{y}(k + 1/k)$$
$$= g[x(k + 1),k] + v_y(k + 1) - g[\hat{x}(k + 1/k),k] \qquad (17.30a)$$
$$= H(k + 1)\hat{e}(k + 1/k) + v_y(k + 1)$$

Substitution of equations (17.28a) and (17.30a) into (17.21) gives:

$$\hat{e}(k+1/k+1) = \phi\hat{e} + \Gamma w - G(k+1)[H(k+1)\hat{e}(k+1/k) + v_y(k+1)] \quad (17.33a)$$

or with index $(k+1)$ replaced by an asterisk:

$$\hat{e}^* = \phi\hat{e} + \Gamma w - G^*H^*\hat{e}(k+1/k) + G^*v_y^* \quad (17.34a)$$

Substitution of equation (17.28a) into (17.34a) gives:

$$\hat{e}^* = \hat{e}(k+1/k) - G^*H^*\hat{e}(k+1/k) + G^*v_y^* \quad (17.34b)$$

Similar to equation (17.37), the covariance matrix P* can be defined as:

$$\begin{aligned} P^* &= E[\hat{e}^*\hat{e}^{*T}] \\ &= (I - G^*H^*)\hat{P}(k+1/k)(I - G^*H^*)^T + G^*R^*G^{*T} \end{aligned} \quad (17.39a)$$

with I the unity matrix.

Define the new J* according to equation (17.43) and minimize $\partial J^*/\partial G^*$. This results in:

$$(I - G^*H)\hat{P}(k+1/k)H^{*T} = G^*R^* \quad (17.49a)$$

or

$$G^* = \hat{P}(k+1/k)H^{*T}[H^*\hat{P}(k+1/k)H^{*T} + R^*]^{-1} \quad (17.50a)$$

with $* = i + 1$.

Equation (17.52) can be written as:

$$\hat{P}(k+1/k) = \phi P\phi^T + \Gamma Q\Gamma^T \quad (17.52a)$$

Substitution of equation (17.50a) into (17.39a) results in:

$$P(k+1/k+1) = [I - G(k+1)H(k+1)]\hat{P}(k+1/k) \quad (17.60)$$

The set of equations obtained now is different from the set in the previous section. The flow diagram for this simplified case is shown in Figure 17.6 (compare to Figure 17.5!).

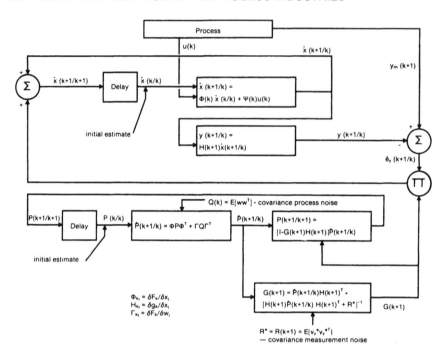

Figure 17.6. Flow diagram of simplified Kalman filter.

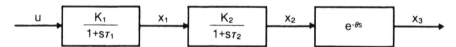

Figure 17.7. Second order/delay system.

ILLUSTRATIONS

Example 1

Consider the simple system shown in Figure 17.7.

Assume that the model equations in discretized form for a certain situation can be written as:

$$\left.\begin{array}{l} x_{1,k+1} = 0.70x_{1,k} + 0.24u_k \\ x_{2,k+1} = 0.80x_{2,k} + 0.16x_{1,k} \\ x_{3,k+1} = x_{2,k} \end{array}\right\} \quad (17.61)$$

The following matrices can therefore be defined:

$$\phi = \begin{pmatrix} 0.70 & 0 & 0 \\ 0.16 & 0.80 & 0 \\ 0 & 1 & 0 \end{pmatrix} \qquad (17.62a)$$

$$\psi = \begin{pmatrix} 0.24 \\ 0 \\ 0 \end{pmatrix} \qquad (17.62b)$$

$$\Gamma = \begin{pmatrix} 1 & 0 & 0 \\ 0 & 1 & 0 \\ 0 & 0 & 1 \end{pmatrix} \qquad (17.62c)$$

and (17.61) can be written as:

$$x_{k+1} = \phi x_k + \psi u_k + \Gamma w_k \qquad (17.63)$$

Assume that only the third state variable is measured, then

$$\left. \begin{aligned} y_{k+1} &= H x_{k+1} \\ H &= (0\ 0\ 1) \end{aligned} \right\} \qquad (17.64)$$

Evaluation of $\hat{P}$ (k + 1/k)

This matrix can be obtained from:

$$\hat{P}(k + 1/k) = \phi P \phi^T + \Gamma Q \Gamma^T \qquad (17.52a)$$

Since Γ is an identity matrix, $\Gamma Q \Gamma^T = Q$, it can easily be seen that:

$$\phi^T = \begin{pmatrix} 0.70 & 0.16 & 0 \\ 0 & 0.80 & 1 \\ 0 & 0 & 0 \end{pmatrix} \qquad (17.65)$$

therefore:

$$\phi P \phi^T = \begin{pmatrix} 0.70 & 0 & 0 \\ 0.16 & 0.80 & 0 \\ 0 & 1 & 0 \end{pmatrix} \begin{pmatrix} P_{11} & P_{12} & P_{13} \\ P_{12} & P_{22} & P_{23} \\ P_{13} & P_{23} & P_{33} \end{pmatrix} \begin{pmatrix} 0.70 & 0.16 & 0 \\ 0 & 0.80 & 1 \\ 0 & 0 & 0 \end{pmatrix} \qquad (17.66)$$

The following is typical for the multiplication of two (2 × 2) matrices:

$$\begin{pmatrix} a_{11} & a_{12} \\ a_{21} & a_{22} \end{pmatrix} \begin{pmatrix} b_{11} & b_{12} \\ b_{21} & b_{22} \end{pmatrix} = \begin{pmatrix} c_{11} & c_{12} \\ c_{21} & c_{22} \end{pmatrix} \qquad (17.67)$$

where

$$\left.\begin{array}{l} c_{11} = a_{11}b_{11} + a_{12}b_{21} \\ c_{12} = a_{11}b_{12} + a_{12}b_{22} \\ c_{21} = a_{21}b_{11} + a_{22}b_{21} \\ c_{22} = a_{21}b_{12} + a_{22}b_{22} \end{array}\right\} \quad (17.68)$$

Evaluation of $\phi P \phi^T$ gives therefore:

$$(17.69)$$

$$\phi P \phi^T = \begin{pmatrix} 0.49P_{11} & 0.112P_{11} + 0.56P_{12} & 0.70P_{12} \\ 0.112P_{11} + 0.56P_{12} & 0.0256P_{11} + 0.256P_{12} + 0.64P_{22} & 0.16P_{12} + 0.8P_{22} \\ 0.70P_{12} & 0.16P_{12} + 0.8P_{22} & P_{22} \end{pmatrix}$$

The following set of equations is now obtained:

$$\left.\begin{array}{l} \hat{P}_{11} = 0.49P_{11,k} + q_{11} \\ \hat{P}_{12} = 0.112P_{11,k} + 0.56P_{12,k} \\ \hat{P}_{13} = 0.70\hat{P}_{12,k} \\ \hat{P}_{22} = 0.0256P_{11,k} + 0.256P_{12,k} + 0.64P_{22,k} + q_{22} \\ \hat{P}_{23} = 0.16P_{12,k} + 0.8P_{22,k} \\ \hat{P}_{33} = P_{22,k} + q_{33} \end{array}\right\} \quad (17.70)$$

in which the process noise covariance matrix

$$\begin{pmatrix} q_{11} & 0 & 0 \\ 0 & q_{22} & 0 \\ 0 & 0 & q_{33} \end{pmatrix} = Q \quad (17.71)$$

Evaluation of the G-Matrix

The G-matrix can be calculated from:

$$G(k+1) = \hat{P}(k+1/k)H(k+1)^T[H(k+1)\hat{P}(k+1/k)H(k+1)^T + R]^{-1} \quad (17.50a)$$

in which $R = E[v_y v_y^T]$, the measurement noise covariance matrix.

Let us evaluate the expression HPH^T first. The following is typical for vector/matrix manipulation:

$$(v_{11} \; v_{12}) \begin{pmatrix} m_{11} & m_{12} \\ m_{21} & m_{22} \end{pmatrix} = (v_{11}m_{11} + v_{12}m_{21} \quad v_{11}m_{12} + v_{12}m_{22}) \quad (17.72)$$

The result of the evaluation is:

$$H\hat{P}H^T = \hat{P}_{33} \tag{17.73}$$

and the inverse in equation (17.50a) becomes $1/(\hat{P}_{33} + r_{33})$.

It can be seen that:

$$\hat{P}(k + 1/k)H(k + 1)^T = \begin{pmatrix} \hat{P}_{13} \\ \hat{P}_{23} \\ \hat{P}_{33} \end{pmatrix} \tag{17.74}$$

thus the G vector becomes:

$$\frac{1}{\hat{P}_{33} + r_{33}} \begin{pmatrix} \hat{P}_{13} \\ \hat{P}_{23} \\ \hat{P}_{33} \end{pmatrix} = \begin{pmatrix} G_{13} \\ G_{23} \\ G_{33} \end{pmatrix} \tag{17.75}$$

Now the new covariance can be evaluated:

$$P(k + 1/k + 1) = [I - G(k + 1)H(k + 1)]\hat{P}(k + 1/k) \tag{17.60}$$

With the aid of (17.75) it can be seen that

$$I - GH = 1 - \frac{\hat{P}_{33}}{\hat{P}_{33} + r_{33}} = 1 - G_{33} \tag{17.76}$$

therefore:

$$P_{ij}(k + 1/k + 1) = [1 - G_{33}(k + 1)]\hat{P}_{ij}(k + 1/k) \tag{17.77}$$

Combination of equation (17.70) and (17.75) gives the equations for the Kalman gain:

$$\left. \begin{aligned} G_{13,k+1} &= \frac{0.70P_{12,k}}{P_{22,k} + q_{33} + r_{33}} \\ G_{23,k+1} &= \frac{0.16P_{12,k} + 0.8P_{22,k}}{P_{22,k} + q_{33} + r_{33}} \\ G_{33,k+1} &= \frac{P_{22,k} + q_{33}}{P_{22,k} + q_{33} + r_{33}} \end{aligned} \right\} \tag{17.78}$$

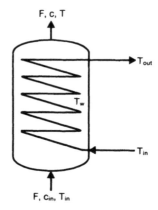

Figure 17.8.   Chemical reactor with heat transfer.

The state estimation for this case is done using equation (17.61):

$$\left.\begin{array}{l}\hat{x}_1(k + 1/k) = 0.70\hat{x}_1(k/k) + 0.24u(k/k)\\ \hat{x}_2(k + 1/k) = 0.80\hat{x}_2(k/k) + 0.16\hat{x}_1(k/k)\\ \hat{x}_3(k + 1/k) = \hat{x}_2(k/k)\end{array}\right\} \quad (17.61)$$

and the new states are calculated using the new measurement:

$$\left.\begin{array}{l}\hat{x}_1(k + 1/k + 1) = \hat{x}_1(k + 1/k) + G_{13}(k + 1)[x_3(k + 1) - \hat{x}_3(k + 1/k)]\\ \hat{x}_2(k + 1/k + 1) = \hat{x}_2(k + 1/k) + G_{23}(k + 1)[x_3(k + 1) - \hat{x}_3(k + 1/k)]\\ \hat{x}_3(k + 1/k + 1) = \hat{x}_3(k + 1/k) + G_{33}(k + 1)[x_3(k + 1) - \hat{x}_3(k + 1/k)]\end{array}\right\} \quad (17.79)$$

Example 2

The previous example served only to illustrate the derivation of the equations and how to apply the Kalman filter algorithm. This example will focus more on the result of the application of the filter. A chemical reactor with heat transfer as shown in Figure 17.8 is considered.

Assuming a first order exothermal reaction and ideal mixing, we can derive the following partial mass balance and energy balance:

$$V\frac{dc}{dt} = Fc_{in} - Fc - Vk_0ce^{-E/RT} \quad (17.80)$$

$$Vpc_p\frac{dT}{dt} = Fpc_p(T_{in} - T) + V\Delta Hk_0ce^{-E/RT} - UA(T - T_w) \quad (17.81)$$

in which V = reactor volume, m³
               c = outlet concentration, kmol/m³

c_{in} = inlet concentration, $kmol/m^3$
F = reactor flow rate, m^3/s
k_0 = kinetic constant, s^{-1}
E/R = ratio of activation energy and gas constant, $°K$
ρc_p = product of density and specific heat, $J/m^3 \, °K$
ΔH = heat of reaction, $J/kmol$
U = overall heat transfer coefficient, $W/m^2 \, °K$
A = heat transfer area, m^v
T_w = average temperature of cooling coil, $°K$
T_{in} = reactor inlet temperature, $°K$
T = reactor outlet temperature, $°K$

It is assumed that it is difficult to analyze the reactor outlet concentration, and, therefore, the Kalman filter will be used to predict the concentration from the (noisy) temperature measurement. Discretizing the model equations using a forward difference approximation and using process data results in the following two difference equations:

$$c_{k+1} = [1 - 1.333F_k - 24 \times 10^7 \, e^{-E_k/RT_k}]c_k + 1.333F_k \qquad (17.82)$$

$$T_{k+1} = [0.896 - 1.333F_k]T_k - 16 \times 10^9 \, e^{-E_k/RT_k}c_k \\ + 1.333F_kT_{in,k} + 0.104T_{w,k} \qquad (17.83)$$

Since the value of the activation energy is not exactly known, E/R is introduced as an additional state variable:

$$E_{k+1}/R = E_k/R + q_{3,k} \qquad (17.84)$$

Introducing the state vector

$$x_k^T = [c_k \; T_k \; E_k/R] \qquad (17.85)$$

and the control vector

$$u_k^T = [F_k \; T_{in,k} \; T_{w,k}] \qquad (17.86)$$

the set of equations can be written in standard form.

The Kalman filter equations as shown in Figure 17.5 were applied recursively, while the inlet temperature was disturbed as shown in Figure 17.9. The estimation of the concentration is shown in Figure 17.10; the estimation of E/R in Figure 17.11. While the process outlet concentration was 0.23 $kmol/m^3$, the initial estimate was 0.30 $kmol/m^3$. After about 40 minutes the estimated concentration tracks the concentration at the process outlet. The activation energy was estimated too high and reached a new value in about 60 minutes. Important is the initial set of estimates as shown in Table 17.1. It is obvious

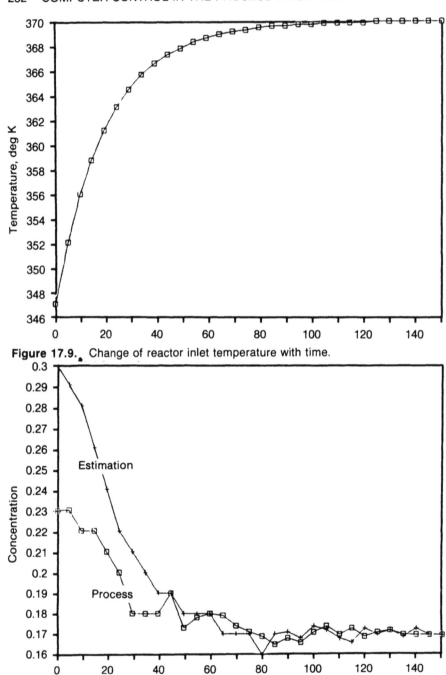

Figure 17.9. Change of reactor inlet temperature with time.

Figure 17.10. Estimation of concentration + = estimation □ = process.

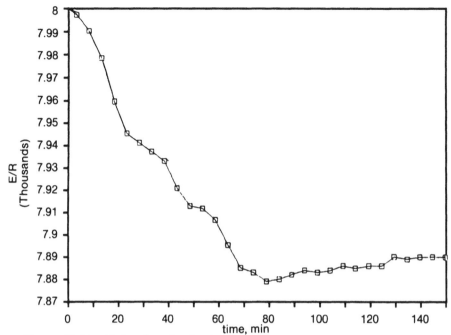

Figure 17.11. Estimation of activation energy.

Table 17.1. Initial Estimate for Kalman Filter

Covariance of the estimation error	
P [1, 1] =	0.1
P [2, 2] =	4.
P [3, 3] =	1000.
Process noise covariance matrix	
Q [1, 1] =	1.
Q [2, 2] =	1.
Q [3, 3] =	10.
Measurement noise covariance	
R [1,1] =	.1

that the initial estimates have an impact on the performance of the filter. Figure 17.12 shows the variation of the Kalman filter gain with time for the correction of E/R. In three time steps the gain reaches a value which is close to the steady state value. Similar results are obtained for the other two gains.

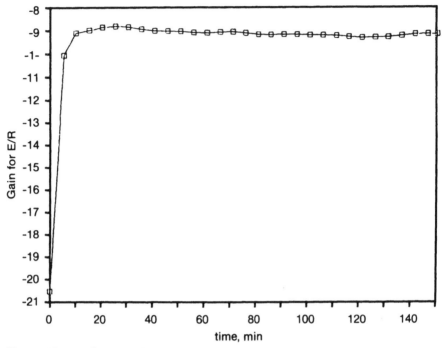

Figure 17.12. Change of Kalman filter gain for E/R with time.

SUMMARY

In this chapter the Kalman filter, which can be interpreted as an optimal state and parameter estimator in the presence of noise, was discussed. It is a useful tool if one wants to predict otherwise unavailable process information from a limited number of measurements. This predicted process information could, if necessary, be used for control.

The Luenberger state observer, as discussed in Chapter 16, is closely related to the Kalman filter; it is the optimal state estimator in the absence of noise. The practical implementation of either technique in a closed loop control scheme requires considerable diagnostic testing in order to ensure that the predicted information is reliable and can be used safely for control.

A similar amount of diagnostic testing is required when implementing the self-tuning controller, discussed in Chapter 15. Here the parameters of a time series model are estimated in an optimal way, and a suitable controller design technique has to be selected in order to derive the control law. If the model is incorrectly identified, the calculated control input will be incorrect. Therefore, considerable care must be taken in identifying the model. Additional checks are necessary on the calculation of the desired control action.

The techniques discussed in Chapters 8–11 have been found to be very powerful, simple, and robust. From an implementation point of view the dynamic reconciliator is probably the easiest to implement and can also be generalized fairly easily to a multivariable case. The multivariable dynamic reconciliator with a linear program can handle dead time, decoupling, and constraints. A major advantage of the techniques presented in Chapters 8-11 is that relatively little diagnostic checking is required to make the control strategy robust and operational. They offer an ideal starting point for the inexperienced control engineer who is eager to improve control but does not want to be bothered by complicated and lengthy control theories.

Literature

1. Giles, R. G., and Bulloch, J. E. Justifying the cost of computer control. *Control Eng.* (May 23, 1976), pp. 49-51.

2. Mellicamp, D. A., Ed. Cache Monograph Series in Real Time Computing. Cambridge, MA: Cache, 1979.

3. Boullart, L., van Cauwenberghe, A., and Delbar, R. L. On-line computers in the process industry — computer hardware. *Polytechnisch Tydschrift* pt/p 35 (1980) nr 6: 348-360; nr 8: 482,490 (in Dutch).

4. McNamara, J. E. *Technical Aspects of Data Communication*. Bedford, MA: Digital Press, 1978.

5. Morris, D. J. *Introduction to Communication, Command and Control Systems*. Oxford: Pergamon Press, 1977.

6. Martinovic, A. Architectures of distributed digital control systems. *Chem. Eng. Prog.* (February 1983), pp. 67-72.

7. Schoeffler, J. D. Distributed computer systems for industrial process control. *IEEE* (February 1984), pp. 11-18.

8. Laduzinsky, A. J. Local area networks expand horizons of control and information flow. *Control Eng.* 30(7):53-56 (1983).

9. Bell, C. G. et al. *Computer Engineering*. Maynard: Digital Corporation, 1978.

10. Tsichritzis, D. C., and Bernstein, Ph. A. *Operating Systems*. New York: Academic Press, 1974.

11. Boullart, L., and Delbar, R. On-line computers in the process industry. *Pt/procestechniek* 35:668-675 (1980), programming techniques (in Dutch); ibid. 36:9-19 (1981), control software (in Dutch).

12. Amrehn, H. Betriebserfabrungen mit Prozesrechern in der chemischen Industrie. *Regelungstechnische Praxis* 1:19–25 (1969).

13. Yourdon, E. *Techniques of Program Structure and Design*. New York: Prentice-Hall, 1975.

14. Munson, J. B. Software maintainability: A practical concern for life cycle costs. *Computer* (November 1981), pp. 103–109.

15. Rony, P. R. Flowcharts, structure charts and written statement programming notations. *Computer Design* 19(3): 172–78 (1980).

16. Beck, W. A., and Carver, T. E. Document your process control system software. *In Tech* (September 1982), pp. 93–95.

17. Shooman, M. L. *Probabilistic Reliability, An Engineering Approach*. McGraw-Hill, 1968.

18. Dolazzo, E. System states analysis and flow graph diagrams in reliability. IEEE Trans on Reliability R-15(3):85–94 (1966).

19. Taylor, D. S. A reliability and comparative analysis of two standby system configurations. *IEEE Trans on Reliability* R-22(1):13–19 (1973).

20. Lynott, F. Control system dependability considerations encompass availability integrity, security. *Control Eng.* (January 1982), p. 22.

21. Amrehn, H. Computer control in the polymerization industry. *Automatica* 13:533–45 (1977).

22. Escher, G., and Weber, H. Einige Auswahlkriterien für den Einsatz von Prozessrechen Systemen in der Chemischen Industrie. *Regelungstechnische Praxis* 3:65–70 (1977).

23. Stübler, H. J. Zuverlassigkeitserfahrungen mit Prozessrechern. *Regelungstechnische Praxis* 6:171–76 (1978).

24. Plogert, K., and Schuler, H. Process control with high reliability data dependant failure detection versus test programs. *Digital Computer Applications to Process Control, Proc. IFAC Conf., 1977* (North Holland Publ. Co., 1977), pp. 695–703.

25. Frey, H. H. Reliability of non-redundant and redundant digital control

computer systems. In *4th IFAC/IFIP Int. Conf. on Digital Computer Applications to Process Control, Zurich, Switzerland, March 19–22, 1974*, pp. 500–513.

26. Roffel, B., and Rijnsdorp, J. E. *Introduction to Process Dynamics, Control and Protection*. Ann Arbor, MI: 1981.

27. Mehta, G. A. The benefits of batch process control. *Chem. Eng. Prog.*(October 1983), pp. 47–52.

28. Zoss, L. M. Control Systems. Applied Instrumentation in the Process Industries, vol. IV,*control systems*. Houston, TX: Gulf Publ. Co., 1979.

29. Ghosh, A. Checklist for batch process computer control. *Chem. Eng.* (Feb. 25, 1980), pp. 88–91.

30. Blickley, G. J. Batch process controls using programmable controllers. *Control Eng.* (July 1984), pp. 81–84.

31. Hanus, R. J. L. A new technique of preventing control windup. *Journal A*:21:1 (1980).

32. Fossard, A. J. *Multivariable System Control*. North Holland Publ. Co., 1977.

33. Roffel, B., Chin, P. A., and Chung, R. D. Control of an in-line blending process. Paper presented at the 36th C.S.Ch.E. Conf., Sarnia, Ont. Canada, October 1986.

34. Jacquot, R. *Modern Digital Control Systems*. Marcel Dekker Inc., 1981.

35. Smith, O. J. M. Closer control of loops with dead time. *Chem. Eng. Prog.* 53(5):217–219 (1957).

36. Bartman, R. V. Dual composition control in a C3/C4 splitter. *Chem. Eng. Prog.* (September 1980), pp. 58–62.

37. Badavas, P. C. Direct synthesis and adaptive controls. *Chem. Eng.* (Feb. 6, 1984), pp. 99–103.

38. Bristol, E. H. On a new measure of interaction for multivariable process control. *IEEE Trans Autom. Control*, January 1986.

39. Shinskey, F. G. *Controlling Multivariable Processes*. ISA, 1981.

40. Alevisakis, G., and Seborg, D. E. An extension of the Smith Predictor method to multivariable linear systems containing time delays. *Int. J. Control*, 3:541–551 (1973).

41. Rijnsdorp, J. E. Chemical Process Systems and Automatic Control. *Chem. Eng. Prog.* 63(7):97 (1967).

42. Westerlund, T. et al. Some simple facts a chemical engineer should know about stochastic control. Technical publication of Åbo Akademi 1985, Report 85-1, Abo, Finland.

43. Duyfjes, G., and Van der Grinten, P. M. E. M. Application of a mathematical model for the control and optimization of a distillation train. *Automatica* 9:537–47 (1973).

44. Harder, H. G., Heller, G. and Laurer, P. R. Optimal control of a chemical rocess by using a computer. *Tech. Mod.* 56:217 (1964).

45. Cutler, C. R. Dynamic matrix control of imbalanced systems. *ISA Transactions* 21(1):1–6 (1982).

46. Marchetti, J. L., Mellichamp, D. A., and Seborg, D. E. Predictive control based on discrete convolution models. *Ind. Eng. Chem. Process Des. Dev. 1983*, 22, 488–495.

47. Asbjornsen, O. A. Feed forward predictive and adaptive control by the dynamic matrix. *Proc. American Control Conference, June 6-8, 1984*, pp. 1864–1869.

48. Cutler, C. R., and Ramaker, B. L. Dynamic matrix control—a computor algorithm. Paper presented at the AICHE 86th national meeting, April 1979.

49. Cutler, C. R. Dynamic matrix control, an optimal multivariable control algorithm with constraints. Ph.D. thesis, University of Houston, Houston, TX, 1983.

50. Richalet, J., Rault, A., Testud, J. L., and Papon, J. Model predictive heuristic control: Applications to industrial processes. *Automatica* 14:413–28 (1978).

51. Martin, G. D. Long range predictive control. *Am. Inst. Chem.* Eng. J. 27(5):748-53 (September 1981).

52. Bray, J. W., High, R. J., Jemmeson, H., and Robson, W. A continuously updated dynamic optimization. *Proc. 3rd IFAC Congress* (London, 1966), p. 8.

53. Roberts, S. M., and Laspe, C. G. On-line computer control of thermal cracking. *Ind. Eng. Chem.* 53:343 (1961).

54. Duncanson, L. A., and Prince, P. C. Interfaces with the process control computer. *Proc. Symp. on Interfaces with the Process Control Computer: The Operator, Engineer and Management, Lafayette, IN, August 3-6, 1971*, p. 86.

55. Shinskey, F. G. *Process Control Systems*. McGraw-Hill, 1967.

56. Bertoni, Corradi, Miami, and Ronconi. Digital control in an electrochemical process. *Proc. 4th IFAC/IFIP Int. Conf. Digital Computer Applications to Process Control, Zurich, Switzerland, March 22-24, 1974, Part 1* (Berlin: Springer Verlag), p. 270.

57. Kerlin, T. W., and Upadhyaya, N. Experiences with identification of power plant dynamics. *7th Triennial World Congr. IFAC: A Link between Science Applications of Automatic Control, Helsinki, Finland, 12-16 June 1978*, (London: Pergamon Press), 105.

58. Halme, A. and Holmberg, A. Application of Control Engineering to Fermentation Process. *7th Triennial IFAC Congr., Helsinki, Finland, 12-16 June 1978*, (London: Pergamon Press), 105.

59. Hammer, H. Praktische Erfahrungen beim Einsatz von Processrechern fur die Rohmehlaufbereitung in der Zementindustrie. *Proc. 4th IFAC/ IFIP Conf. Zurich, Switzerland, March 19-22, 1974, Part 1*, (Berlin: Springer Verlag), p. 410.

60. Sahata, M. Standardized analysis and open loop control of complex systems by use of process computers. *Regelungstechn and Process Datenverarb* 21(1):9 (1973).

61. Swenber, G. Practical data reconciliation methods for industrial measurements. *De Ingenieur* (May 28, 1971), p. 65.

62. Crowther, R. H., Pitrak, J. E., and Ply, E. N. Computer control at American Oil. *Chem. Eng. Prog.* 57(6):39 (1961).

63. Eykhoff, P. Identification and system parameter estimation. *Proc. 3rd IFAC Symp., The Hague/Delft, June 12–15, 1973; N. S. Rajbman. Proc. 4th IFAC Symposium, Tbilisi (USSR), Sept 21–27, 1976*, (North Holland Publ. Co.)

64. Eykhoff, P. *System Identification: Parameter and State Estimation.* John Wiley & Sons, 1974.

65. Crico, A. Automatic control of distillation columns with several side streams. *Genie Chimique* 1:12 (1979).

66. Bridgewater, K. et al. Digital computer control at a petroleum refinery. Paper presented at Proc. 3rd IFAC Congr., London, 1966.

67. Pavlik, E. Observation theory in process automation. Siemens Forschungsund Entwicklungsberichte Bd. 9 (1980), No. 5.

68. Luenberger, D. G. An introduction to observers. *IEEE Trans AC 16* (1971), pp. 596–602.

69. Seborg, D. E., Fisher, D. G., and Hamilton, J. C. An experimental evaluation in multivariable control systems. *Proc. 4th IFAC/IFIP Conf. Zurich, Switzerland, March 1974*, p. 144.

70. Church, D. F. Current and projected pulp and paper industry problems in process control modelling. Am. Inst. Chem. Eng. Symp. Ser. 72 (1976), p. 19.©

71. Powell, D. et al. A computer coordinated bleach plant *Instr. Tech.* (May 1974), p. 42.

72. Amrehn, H. Computer control in the polymerization industry Automatica 13:533 (1977).

73. Lafayette, W. Use of steady state techniques in Industrial Control. *Proc. Adv. Annual Control Conf. W., Lafayette, IN, 1 April 1979*, p. 9.

74. Kennedy, J. P. *The Use of Process Computers as Industrial Robots for Refining Operation* (ISA, 1983), p. 77.

75. Bronfenbrenner, J. C., and Touchstone, A. T. Inferred product compositions for improved distillation control. 90th Am. Inst. Chem. Eng. national meeting, Houston, Texas, April 5-9, 1981.

76. Jain, G. P., Gains, L. D., and Wainwright, D. E. Computer control of crude towers. *Oil Gas J.* (Dec. 10, 1979).

77. Richalet, J. et al. Model algorithmic control of industrial processes. *Proc. 5th IFAC/IFIP Conference on Digital Computer Applications to Process Control, The Hague, June 14-17, 1977.*

78. Baxley, R.A. The treatment of distillation systems in on-line optimization. *Proc. 5th IFAC/IFIP Conference on Digital Computer Applications to Process Control, The Hague, June 14-17, 1977.*

79. Latour, P.R. Objectives and strategies for composition control of distillation columns. *Instr. Tech.* 25:7,67 (1978).

80. Lamb, M. Y. Computer control of a propylene upgrading unit. 85th Am. Inst. Chem. Eng. national meeting, Philadelphia, June 1978.

81. Kennedy, J. P. New control package is aid to cat reforming. *Oil Gas J.* (Sept. 24, 1979), pp. 24-35.

82. Webb, P. U., Stewart, W. S., and Smith, D. E. Process computer applications for control of refinery units. *Proc. Nat. Petr. Ref. Ass. Comp. Conf., New Orleans, LA, Nov 15-17, 1976.*

83. Carr, N. L., Kramer, S. J., and Stahlfield, D. L. U.S. Patent No. 3725653 (April 3, 1973).

84. Sayles, J. H. et al. Computer control maximizes hydrocracker throughput. *Instr. Tech.* (May 1973), pp. 61-70.

85. Sourander, M. L. et al. Control and Optimization of Olefin Cracking Heaters (Hydrocarbon Processing, June 1984), pp. 63-69.

86. Dy Liacco, T. E. Digital computer applications in the control of electrical power systems. *Proc. 5th IFAC/IFIP Conf., The Hague, June 14-17, 1977.*

87. Laspe, C. G. Recent experiences in on-line optimizing control of industrial processes. *Proc. Annual Adv. Control Conf., W. Lafayette, IN, 1979,* pp. 175-87.

88. *Special Report: Advanced Process Control Handbook*. Hydrocarbon Processing, February 1986, vol. 65, no. 2.

89. Aström, K. J., and Wittenmark, B. On self-tuning regulators. *Automatica* 9:105–99 (1973).

90. Clarke, D. W., and Gawthrop, B. A. Self-tuning controller. Proc. Inst. of Electrical Engineers, 1975, 122, pp. 929–34.

91. Aström, K. J. Theory and applications of adaptive control – a survey. *Automatica* 19(5):471–86 (1973).

92. Seborg, D. E., Shah S. L., and Edgar, T. F. Adaptive control strategies for process control: A survey. Am. Inst. Chem. Eng. Diamond Jubilee meeting, Washington, DC, Nov 1983.

93. Unbehauen, H. Theory and applications of adaptive control. 7th IFAC/IFIP/IMAC Conference on Digital Computer Applications to Process Control, Vienna, 17–20 Sept 1985.

94. Wittenmark, B., and Aström, K. J. Practical issues in the implementation of self-tuning control. *Automatica* 20(5): 595–605 (1984).

95. Harris, T. J., MacGregor, J. F., and Wright, J. D. Self-tuning and adaptive controllers: An application to catalytic reactor control. *Technometrics* 22(2):153–64 (1980).

96. MacGregor, J. F., Harris, T. J., and Wright, J. D. Duality between the control of processes subject to randomly occuring deterministic disturbances and ARIMA stochastic disturbances. *Technometrics* 26(4):389–97 (1984).

97. Box, G. E. P., Jenkins, G. M., and MacGregor, J. F. Some recent advances in forecasting and control, part II. *J. Roy. Static. Soc.*, C23, 158 (1974).

98. Ydstie, B. E., Kershenbaum, L. S., and Sargent, R. W. H. Theory and application of an extended horizon self-tuning controller. *Am. Inst. Chem. Eng. J.* 31(11):1771–80 (1985).

99. Goodwin, J. C., and Payne, R. L. *Dynamic System Identification: Experimental Design and Data Analysis* (Academic Press, 1977), pp. 177–81.

100. Kwakernaak, H., and Sivan, R. *Linear Optimal Control Systems*. John Wiley & Sons, 1972.

101. Sage, A. P., and Melsa, J. L. *Estimation Theory with Application to Communication and Control*. McGraw-Hill, 1971.

102. Gelb, A. *Applied Optimal Estimation*. MIT, 1974.

103. Luenberger, D. G. Observing the state of a linear system. *IEEE Trans MI 1-8, 1964, no. 2*, pp. 74–80.

104. Luenberger, D. G. Observers for multivariable systems. *IEEE Trans. AC11, 1966, no. 2*, pp. 190–97.

105. Kalman, R. E. A new approach to linear filtering and prediction problems. Trans. A.S.M.E. Ser. D. *J. Basic Eng.* 82:35–45 (1960).

106. Kalman, R. E., and Bucy, R. S. New results in linear filtering and prediction theory. Trans. A.S.M.E. Ser. D. *J. Basic Eng.* 83:95–108 (1961).

107. Jazwinski, A. H. *Stochastic Processes and Linear Filtering Theory*. Academic Press, 1970.

108. Brown, R. G. *Introduction to Random Signal Analysis and Kalman Filtering*. John Wiley & Sons, 1983.

109. Kipperissides, C., MacGregor, J. E., and Hamielec, A. E. Suboptimal stochastic control of a continuous latex reactor. *Am. Inst. Chem. Eng. J.* 27:13–19 (1981).

110. Jo, J. H., and Bankoff, S. G. Digital monitoring and estimation of polymerization reactors. *Am. Inst. Chem. Eng. J.* 22(2):361–368 (1976).

111. Schuler, H. Estimation of states in a polymerization reactor. *Proc. IFAC Conf. on Automation, Ghent, Belgium, 1980*, pp. 369–376.

112. Canavas, C. State and parameter estimation through non-linear filtering – an application to a fixed bed reactor. *Proc. Telecon 1984 (IASTED Int. Conf.), August 27–30, 1984, Halkidiki, Greece*.

Nomenclature

a	white noise or constant
A	max (θ, τ) as defined in equation (8.9) also used for area, availability, and as a matrix.
A_o	bias term as defined in equation (10.23)
δA	change in analyzer reading
b	constant
B	min (θ, τ) as defined in equation (8.10), also used as a matrix
c	concentration, kg/m^3
C	conversion, also used as a matrix
c_p	specific heat, J/kg K
c_{index}	component value, \$/kg
d	plant/model mismatch, also used as parameter
e	setpoint − measurement, also used as parameter
e_y	innovation, measurement − prediction
E	activation energy
E[ ]	expectation
ER	external reflux, kg/s
f	dead time
f( )	functional relationship used in state space models
F	feed flow or factor defined in Table 12.1
F( )	functional relationship used in discrete models for state variables
g()	functional relationship for output variables
G	transfer function or matrix
h	heat transfer coefficient, W/m^2K
H	$\delta g/\delta x$ as defined in Chapter 17, also used as a matrix of gain coefficients in Chapter 11

247

H (k) $y_k - Y_{sp,k-1}$ as defined in equation (12.6)
ΔH   heat of reaction, J/kg

I   identity matrix
IR   internal reflux

J   generally used for objective function, also used in Chapter 16
               for Jordan canonical matrix

k   time
K, K*   gain
K_p process gain
K_c controller gain
KLAG $e^{\Delta t/\tau} \simeq \tau/(\tau + \Delta t)$, lag factor
KLEAD $K\tau_1/(\tau_2 + \Delta t)$, lead factor

l_u process gain as defined by equation (11.1)
l_y process gain as defined by equation (11.2)
LMTD   logarithmic mean temperature difference, K

Δm $y - y_{max}$, used in constraint control
M   mass, kg
MRT, MTTR   mean repair time, λ^{-1}, yr^{-1}
MTBF   mean time between failures, MTTF & MRT, yr
MTTF   mean time to failure, μ^{-1}, yr^{-1}

N $E[v_u v_u{}^T]$, input measurement noise covariance matrix
N_k disturbance

P $E[\hat{e} \hat{e}^T]$, covariance matrix of the estimation error, also used as
               probability and pressure (kPa)

q   off-line measured output variable
Q $E[ww^T]$, covariance matrix of the disturbance
Q_h heat input, W/m^2

r   rate of reaction, $mol/m^3 s$
R $E[v_y v_y{}^T]$, output measurement noise convariance matrix
               also used as reflux, kg/s, and gas constant

s   Laplace operator
S_H standard deviation of process H
S   separation factor according to equation (13.11)

t	time, min, s
Δt	discretization interval, min, s
T	temperature, K
ΔT	temperature difference, K
u	controller output
u_d	decoupler output
U	overall heat transfer coefficient, W/m^2K
v	measurement noise
V	volume, m^3
w	process noise, disturbance
x	sate variable
x_B	bottom composition
x_D	overhead composition
x_{BO}	bottom bias term, defined in equation (11.52)
x_{DO}	overhead bias term, defined in equation (11.51)
y	process output
y_o	bias term for process, output according to equation (11.67)
z^{-1}	backward shift operator, defined by equation (7.25)

Greek symbols

α	control law/model parameter
β	control law/model parameter
$\beta*$	control law/model parameter
Γ	$\partial F/\partial w$, the derivative of the process model with respect to the disturbances
δ	variation around the steady state, also used as model parameter
Δ	$1-z^{-1}$ operator
ϵ	forecast error

ζ	parameter to limit changes in process input
θ	dead time
$\theta(z^{-1})$	polynomial in z^{-1} used in disturbance model
$\varkappa$	noise model parameter
λ	weighting coefficient as defined in equation (12.7)
	also used as Lagrangian multiplier and as repair rate ($=$ total timed in use/average repair time)
μ	failure rate, number of failures/year
ϱ	density, kg/m³
τ_1, τ_2	lead and lag times used in lead-lag element
τ_R	time constant for desired closed loop response
τ_P	process time constant
τ_i	integral time constant
τ_d	derivative time constant
∇	$1-z^{-1}$ operator
ϕ	function of process variable x; also used for $\partial F/\partial x$, the derivative of the process model with respect to the state variables
$\phi\,(z^{-1})$	polynomial in z^{-1} used in disturbance model
ψ	$\partial F/\partial u$, the derivative of the process model with respect to the control inputs
$\psi\,(z^{-1})$	polynomial in z^{-1} used in disturbance model
ω	model parameter
Ω	$\partial g/\partial u$, the derivative of the measurement model with respect to the control inputs.

Subscripts

d,D	decoupler
df	delayed and filtered
ff	feed-forward
i	input

m	maximum or model
o	output
oh	overhead
p	process
sp	setpoint
ss	steady state

Superscripts

T	transposed
^	estimated value
ˇ	predicted value

INDEX